Geography of the Physical Environment

About this Series

The Geography of the Physical Environment book series provides a platform for scientific contributions in the field of Physical Geography and its subdisciplines. It publishes a broad portfolio of scientific books covering case studies, theoretical and applied approaches as well as novel developments and techniques in the field. The scope is not limited to a certain spatial scale and can cover local and regional to continental and global facets. Books with strong regional focus should be well illustrated including significant maps and meaningful figures to be potentially used as field guides and standard references for the respective area.

The series appeals to scientists and students in the field of geography as well as regional scientists, landscape planners, policy makers, and everyone interested in wide-ranging aspects of modern Physical Geography. Peer-reviewed research monographs, edited volumes, advance and under-graduate level textbooks, and conference proceedings covering the major topics in Physical Geography are included in the series. Submissions to the Book Series are also invited on the theme 'The Physical Geography of...', with a relevant subtitle of the author's/editor's choice.

More information about this series at http://www.springer.com/series/15117

Levan Tielidze

Glaciers of Georgia

 Springer

Levan Tielidze
Institute of Geography
Tbilisi State University
Tbilisi
Georgia

Some materials were already published just in Georgia, author owns copyright

ISSN 2366-8865   ISSN 2366-8873   (electronic)
Geography of the Physical Environment
ISBN 978-3-319-50570-1   ISBN 978-3-319-50571-8   (eBook)
DOI 10.1007/978-3-319-50571-8

Library of Congress Control Number: 2016960192

Cover image by Sonja Weber, München

Printed on acid-free paper

This Springer imprint is published by Springer Nature
The registered company is Springer International Publishing AG
The registered company address is: Gewerbestrasse 11, 6330 Cham, Switzerland

Preface

The monograph is based on the past several-year research results, which were obtained during the study of modern and old glaciations of the Georgian Caucasus. As a result of these surveys, the latest materials on the modern glaciers' morphology, morphometry, and dynamics have been obtained, as well as on structure of moraines and the river terraces, geodynamics of the relief, snow and firn lines location.

At various times, we conducted field surveys in almost every glacier basins in the southern and northern slopes of the Georgian Caucasus. Apart from the field researches, we used the remote sensing method. After processing the latest aerial images (Landsat L5, L8 OLI, ASTER) by modern computer programs (ArcGIS, ENVI, Google Earth), we got the quite accurate information about the glaciers difficult to access. This mainly refers to the glaciers that are located in the temporarily occupied Abkhazeti and Tskhinvali region, where it is so far impossible for us to conduct field research. During our research we also used the traditional methods: glacio-geomorphological, cartographical, aerial image processing, and petrographic.

The monograph includes a number of new statements and conclusions; among them the following are essential:

1. Principally new numerical and qualitative characteristics of present glaciers and their dynamics have been derived and full databases of Georgia's modern glaciation have been composed;
2. Valley glaciers fluctuation synchronicity has been revealed after the Little Ice Age (LIA) maximum;
3. Reconstruction of the Late Pleistocene (Wurmian) and Holocene glaciations has been investigated. The maps of the distribution of the Late Pleistocene glaciation of the Georgian Caucasus have been compiled.

The main theoretical statements and conclusions have been developed in the Vakhushti Bagrationi Institute of Geography in Georgia. During 2014–2015, certain part of the research has been performed in the United States of America, in the Glaciology and Remote Sensing Laboratory of the Climate Change Institute of the University of Maine, also, during 2015–2016—in Canada, at the University of Northern British Columbia.

Data obtained on present state and dynamics of the glaciers of Georgia can be used for water supply and development of hydropower in the settlements of mountainous areas. Quantitative data obtained on the present state of the nival–glacial system is necessary for the design and construction of the tourist-recreational objects in the high mountain zone, as well as for the development of tourism and alpinism.

We hope that this monograph will be of great assistance for the public who is interested in any information about glaciers of Georgia.

Special thanks from the author to Ms. Nino Chikhradze for literary editing, translating, and cooperation during the preparation of this book.

The authors of the photos used in this book: Levan Tielidze, Ramin Gobejishvili Davit Tsereteli, K.P. Rototaeva, and Giorgi Gotsiridze. The photos of the glaciers made by the Italian photographers Vittorio Sella and Mor Von Dechy in the nineteenth century and the nineteenth–twentieth-century photos of the glaciers kept in the fund of the Museum of Geography at Tbilisi State University are also used.

Author will accept all the topic-related comments with gratitude.

Tbilisi, Georgia   Levan Tielidze

Contents

About the Author

Levan Tielidze Ph.D. in Glaciology and Geomorphology at Ivane Java-khishvili Tbilisi State University, Georgia. He is Researcher-Scientist of the Vakhushti Bagrationi Institute of Geography.

His research field is the glacio-geomorphological study of the mountain glaciers and their mapping using geo-information systems (GIS). He studies the dynamics of the glaciers against the background of climate change using remote sensing. He also studies the Little Ice Age (LIA) glaciation and the reconstruction of the glaciation in the Late Pleistocene and Holocene.

During 2014–2015 he passed the internship in the United States of America, in Climate Change Institute (University of Maine) and during 2015–2016—in Canada, University of Northern British Columbia. He is the author of about 40 scientific papers including six monographs.

The Historical Overview

Research of glaciers has a long history in the Caucasus. Great Georgian scientist Vakhushti Bagrationi gives the first scientific information on the glaciers of Georgia in the beginning of the eighteenth century ["There are big mountains, which have the Caucasus to the North from the Black Sea to the Caspian Sea, the height of which is of one day walking and the highest of it is permanently frosty, the length of the ice is of k-l arm, and in summer it breaks and, if a man stays there, he cannot endure the cold even for a short time; and the rivers flow under it, and the ice is green and red, as a rock due to its age"] (Vakhushti 1941).

After almost hundred years the foreign scientists began to describe the glaciers of Georgia. Information about the glaciers of Georgia can be found in the works of Abich (1865), Freshfield (1896), Radde (1873), Dinik (1890), Rashevskiy (1904), etc.

One of the first foreign travelers, who visited the Caucasus in the nineteenth century, was Douglas Freshfield (1869). He wrote in the account of a visit in 1868 that the Caucasus was lesser known than the Andes and Himalayas. Merzhbacher (1901, cited Horvath 1975) found evidence of glacier advances in recent centuries in a valley of the central Caucasus. Ruined buildings lie close to glacier tongue and local legends and songs told of a glacier near Ushguli (the mountain village in Georgia), probably the Khalde glacier, then six miles away from the village, having advanced and destroyed all of it but for the church. The people still held an annual festival in thanksgiving (Grove 1988), etc. All this information greatly assisted us in determining the dynamics of the individual glaciers.

In the years 1880–1910 the topographical surveying of the Greater Caucasus was carried out. On the basis of the created maps, Podozerskiy (1911) compiled the first detailed catalog of the glaciers, which still has not lost its importance, but it must be mentioned that the errors were made during its compilation. Reinhardt (1916, 1936, 1937) noted these errors further, who compiled the new catalog for many glacial basins of the investigated region and defined the location of the snow line. The research conducted by A. Reinhardt is of high quality and more reliable by its scientific value in comparison with its previous researchers.

Interesting researches were conducted by Rutkovskaya (1936) in connection with the 2nd International Polar Year. During 1932–1933, the glaciation of the Enguri River was studied and the dynamics (in the one-year period) of the individual glaciers were identified.

In 1959 P.A. Ivankov gave us the total number and area of glaciers of the study area based on the new topographic maps and the aeroimages of 1946. In the same year Kovalev (1961) described the glaciers in detail and carried out their labeling.

Much work has been conducted by D. Tsereteli for the study of the glaciers of Georgia, who in 1937 together with Al. Aslanikashvili surveyed several glaciers and in 1963 gave us the dynamics of the glaciers during the period 1937–1960.

Particularly should be mentioned the great and versatile work, which was done by the Glaciological Laboratory of Vakhushti Bagrationi Institute of Geography, the multiannual work of which is summarized in the 1975 year's edition of the Catalog of Glaciers, as well as by the Hydrographical Division of the Hydro-Meteorological Institute, which published the work about the Glaciers of the Greater Caucasus (Editors: V. Tsomaia and E. Drobishev, 1970).

The many years research of various glaciers in the major river basins by R. Gobejishvili should also be noted. It can be considered his honor that after the 1990s the glaciological studies have not been stopped in Georgia.

L. Maruashvili, D. Ukleba. T. Kikalishvili, G. Kurdghelaidze, D. Tabidze, R. Khazaradze, O. Nikolaishvili, V. Tsomaia, O. Drobishev, R. Shengelia, R. Gobejishvili, K. Mgeladze, T. Lashkhi, Sh. Inashvili, N. Golodovskaia, L. Serebruanny, A. Orlov, O. Nadirashvili, N. Zakarashvili, A. Rekhviashvili, O. Samadbegishvili and others studied the glaciers of Georgia according to the river basins.

Glacial-geomorphological works were being carried out from 1968 (R. Gobejishvili). The largest glaciers of the different river basins were surveyed by the phototheodolite method by: Zopkhito-Laboda, Kirtisho, Brili, Chasakhtomi, Edena, Khvargula, Boko, Buba, Tbilisa, Adishi, Chalaati, Dolra, Kvishi, Ladevali, Shkhara, Namkvani, Koruldashi, Marukhi, Klichi and the cirque type glaciers of the Klichi basin.

Finally, we would like to say that since 1930s the observation on the Western, Central and Eastern Caucasus glaciers in Georgia has a nearly continuous nature. Glaciology group of the Vakhushti Bagrationi Institute of Geography is still conducting the constant monitoring of the glaciers of Abkhazeti, Svaneti, Racha and Kazbegi Caucasus.

References

Abich H (1865) Isledovanie nastaiashix i drevnix lednikov Kavkaza (The study of current and ancient glaciers of the Caucasus). Cb, cvedenii o Kavkaze. T. 1. Tiflis. (in Russian)

Dinnik NY (1890) Present and old glaciers of Georgia. KOIRGO 14, 88–102, Tiflis (in Russian)

Grove JM (1988) The little ice age. Book. ISBN 0-415-01449-2. Metchuen and Co. Ltd. London

Horvath E, Field WO (1975) The Caucasus. In: Field WO (ed) Mountain glaciers of the northern hemisphere. Cold Regions Research and Engineering Laboratory. Hanover, NH

Kovalev PV (1961) Sovremennie i drevnie oledenenie bacceina r. Inguri (Modern and ancient glaciation Enguri River basin). Mat. Kavkaz. Eksped. T. 2. Xarkov. Izd XGU (in Russian)

Podozerskiy KI (1911) Ledniki Kavkazskogo Khrebta (Glaciers of the Caucasus Range). Zapiski Kavkazskogo otdela Russkogo Geograficheskogo Obshchestva, Publ. Zap. KORGO., Tiflis, 29, 1, p 200 (in Russian)

Radde GI (1873) Predvoritelnii otchet o puteshestvii d-ra Radde po Kavkazu (Preliminary report on the journey of Dr. Radde the Caucasus.). Zap KORGO, kn. 8. (in Russian)

Rashevskiy NN (1904) Cherez Gebivcek (Through pass Gebi). ERGO. T. 3. (in Russian)

Reinhardt AL (1916) Snejnaya granica Kavkaze (The snow line in the Caucasus). Izvestia Kavkazskogo otdela Imperatorskogo Russkogo Geograficheskogo Obshchestva, T. 24. vol 3 (in Russian)

Rutkovskaya VA (1936) Sections: Upper Svaneti Glaciers, pp 404–448. In: Transactions of the glacial expeditions. Vol 5. Caucasus, the glacier regions. USSR Committee of the II international polar at the centr. adm. of the hydro-meteorological service. Leningrad

Tsomaia VS, Drobyshev OA (1970) The results of glaciological observations on the glaciers of the Caucasus. ZakNIGMI, no. 45 (in Russian)

Vakhushti (1941) Description of Georgian Kingdom (Geography of Georgia). Tbilisi State University Press. Tbilisi (in Georgian)

Introduction

1

Abstract

The location of glaciological regions of the Caucasus according to Global Land Ice Measurements from Space (GLIMS) and Randolph Glacier Inventory (RGI) is represented in this chapter. Division of the Caucasus Mountains by separate units is also considered. The climate and relief of the Caucasus is described. The significance of glaciers as for Georgia as well as for other mountainous regions of the Caucasus is reviewed generally.

Keywords

The Caucasus · Randolph Glacier Inventory (RGI) · Global Land Ice Measurements from Space (GLIMS) · The Lesser Caucasus

The glaciers are indivisible part of the environment and are a good indicator of the past and current climate change (Tielidze et al. 2015). Alpine glaciers are an important component of the global hydrologic cycle. Glaciers can help to regulate stream flows in regions where water is stored during cold wet times of the year and later released as melt water runoff during warm dry conditions (Beniston 2003; Earl and Gardner 2016). The most serious impact of vanishing mountain glaciers undoubtedly concerns the water cycle from regional to global scales. Glacier melting will probably dominate sea level rise during our century (Meier et al. 2007).

Distribution and diversity of glaciers on the Earth determine their grouping in separate regions by foreseen of the external conditions of existence of glaciers. Such zoning allows us to better understand the characteristics of glaciers' regime and the synchronism of their action in different regions, as well as to relate the distribution of glaciers to the general circulation of the atmosphere and the relief orography.

Nineteen regions have been distinguished on the Earth based on Randolph Glacier Inventory (RGI) (Pfeffer et al. 2014), which is intended for the estimation of total ice volumes and glacier mass changes at global and large regional scales. It is supplemental to the Global Land Ice Measurements from Space initiative (GLIMS). Production of the RGI was motivated by the preparation of the Fifth Assessment Report of the Intergovernmental Panel on Climate Change (IPCC AR5) (GLIMS Technical Report). As a

© Springer International Publishing AG 2017
L. Tielidze, *Glaciers of Georgia*, Geography of the Physical Environment,
DOI 10.1007/978-3-319-50571-8_1

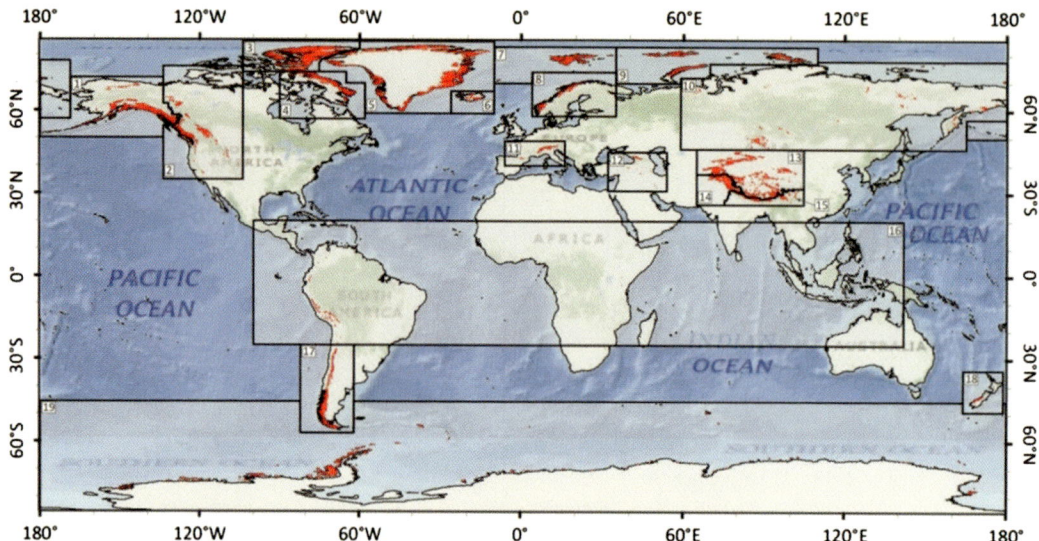

Fig. 1.1 First-order regions of the Randolph Glacier Inventory (version 4.0). *1.* Alaska; *2.* Western Canada and US; *3.* Arctic Canada North; *4.* Arctic Canada South; *5.* Greenland Periphery; *6.* Iceland; *7.* Svalbard; *8.* Scandinavia; *9.* Russian Arctic; *10.* North Asia; *11.* Central Europe; *12.* **Caucasus and Middle East**; *13.* Central Asia; *14.* South Asia West; *15.* South Asia East; *16.* Low Latitudes; *17.* Southern Andes; *18.* New Zealand; *19.* Antarctic and Subantarctic (Pfeffer et al. 2014)

result of the mentioned inventory, the Caucasus is presented together with the Middle East (as one region) (Fig. 1.1), but the Caucasus is much larger than the Middle East by the modern glaciation size and it will be interesting if we consider it as a separate region in our work.

The Caucasus Mountains are aligned west-northwest to east-southeast between 40–44° N and 40–49° E and span the borders of Russia, Georgia, Armenia, and Azerbaijan. They consist of two separate mountain systems: **the Greater Caucasus** extends for ∼1300 km between the Black Sea and Caspian Sea, whilst **the Lesser Caucasus** runs parallel but approximately 100 km to the south. The Caucasus Mountains originate from collision between the Arabian plate to the south and the Eurasian plate to the north and the region is tectonically active with numerous small earthquakes (Stokes 2011).

According to location the Greater Caucasus is divided into three parts: Western, Central, and Eastern. The borderline among them runs near the meridians of the Mount Elbrus (5642 m) and the Mount Kazbegi (5033 m). In the mountainous system of Caucasus the highest is the Central Caucasus. Several peaks are higher than 5000 m (e.g. Elbrus, Dikhtau, Shkhara massif, and Kazbegi). It is in this section the Europe's highest peak Elbrus (5642 m) with its glacial complex.

The Caucasus Mountains are characterized by strong longitudinal gradients that produce a maritime climate in the west and a more continental climate in the east. Trends in precipitation, for example, reveal that westernmost areas typically receive around three to four times as much as eastern areas (Horvath and Field 1975). The southern slopes are also characterized by higher temperatures and precipitation, which can be up to 3000–4000 mm in the southwest (Volodicheva 2002). Much of this precipitation falls as snow, especially on windward slopes of the western Greater Caucasus, which are subjected to moist air masses sourced from the Black Sea (Stokes 2011).

According to the conditions of relief, the northern slope of the Caucasus is more favorable for formation of glaciers than the southern one. This is contributed by high hypsometry and extremely partitioned slopes, gorges, and depressions, represented by wide cirques of Wurm period.

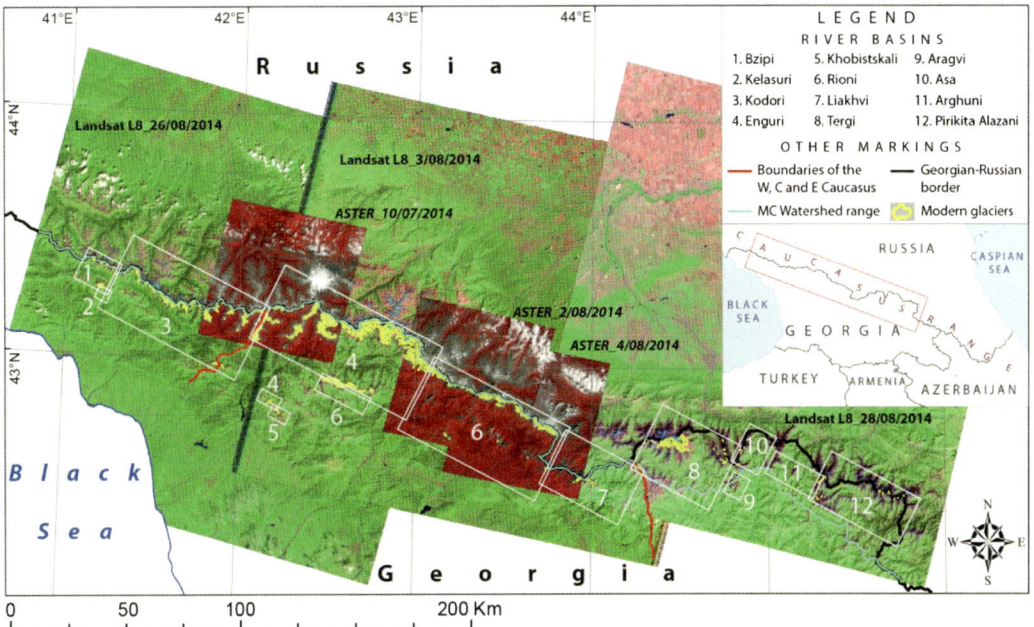

Fig. 1.2 Georgian Caucasus glacier outlines (in *yellow*) derived from Landsat 8 and ASTER imagery. *White rims* show the individual river basins

In the Caucasus the current number of glaciation is ~2000 with a total area of ~1100 km² and volume ~68 km³ (Radić et al. 2014). Approximately 33% of the glaciers of the Caucasus is located in Georgia (Fig. 1.2). These Glaciers are an important source of water for agricultural production in Georgia, and runoff in large glacially fed rivers (Kodori, Enguri, Rioni, Tskhenistskali, and Nenskra) supplies several hydroelectric power stations. Glacial melt waters are one of the main factors in river runoff formation in the mountainous areas of Georgia. It is necessary to know Glacial waters daily volatility for mountaineering, tourism and mountainous areas of livestock and other sectors of operation. Glacier melt water is also important in terms of water supply in the mountainous regions of Georgia. In the mountainous regions (Svaneti, Kazbegi, Racha, and Abkhazeti), in addition to the tourist—recreational purposes, a great role owned the glacial landscapes in the development of the recreational facilities.

Also glacier outburst floods and related debris flows are a significant hazard in Georgia and in the Caucasus (Bogatikov et al. 2003). Unfortunately, such hazards are relatively common in this region and have led to major loss of life. In September 20 of 2002, for example, Kolka Glacier (North Ossetia) catastrophic ice-debris flow killed over 100 people (Evans et al. 2009), and in May 17 of 2014, Devdoraki Glacier (Georgia) catastrophic rock–ice avalanche and glacial mudflow killed nine people. Future trends in glaciers variations are thus a topic of considerable interest to the region.

References

Beniston M (2003) Climatic change in mountain regions: a review of possible impacts. Clim Change 59:5–31

Bogatikov OA, et al (2003) Catastrophic rock-ice collapse and rapid shove of the Kukurtli Glacier (Elbrus Volcano, Northern Caucasus) in the 1st and 2nd centuries. Rep Earth Sci 391(5):627–630

Earl L, Gardner A (2016) A satellite-derived glacier inventory for North Asia. Ann Glaciol 57(71). doi:10.3189/2016AoG71A00850

Evans SG, Tutubalina OV, Drobyshev VN, Chernomorets SS, McDougall S, Petrakov DA, Hungr O (2009) Catastrophic detachment and high-velocity long-runout flow of Kolka Glacier, Caucasus Mountains, Russia in 2002. Geomorphology 105: 314–321

Horvath E, Field WO (1975) The Caucasus. In: Field WO (ed) Mountain Glaciers of the Northern Hemisphere. Cold Reg Res Eng Lab, Hanover

Meier MF, Dyurgerov MB, Rick UK, O'Neel S, Pfeffer WT, Anderson RS, Anderson SP, Glazovsky AF (2007) Glaciers dominate eustatic sea-level rise in the 21st century. Science 317:1064–1067. doi:10.1126/science.1143906

Pfeffer WT, Arendt AA, Bliss A, Bolch T, Cogley JG, Gardner AS, Hagen J, Hock R, Kaser G, Kienholz C, Miles ES, Moholdt G, Mölg N, Paul F, Radic V, Rastner P, Raup BH, Rich J, Sharp MJ, The Randolph Consortium (2014) The Randolph Glacier inventory: a globally complete inventory of glaciers. J Glaciol 60:537–552. doi:10.3189/2014JoG13J176

Radić V, Bliss A, Beedlow AC, Hock R, Miles E, Cogley JG (2014) Regional and global projections of twenty-first century glacier mass changes in response to climate scenarios from global climate models. Clim Dyn 42(1–2):37–58. doi:10.1007/s00382-013-1719-7

Stokes CR (2011) Sections: Caucasus Mountains. In: Encyclopedia of snow, ice and glaciers. Springer, Dordrecht, pp 803–808

Tielidze LG, Kumladze R, Asanidze L (2015) Glaciers reduction and climate change impact over the last one century in the Mulkhura River Basin, Caucasus Mountains, Georgia. Int J Geosci 6:465–472. doi:org/10.4236/ijg.2015.65036

Volodicheva N (2002) The Caucasus. In: Shahgedanova M (ed) The physical geography of Northern Eurasia. Oxford University Press, Oxford, pp 350–376

Part I

Modern Glaciers of Georgia

Abstract

This chapter includes the relief and climatic factors of glaciers' generation in Georgia. There are described the individual sectors of Caucasus and also those branch ranges, which are related to the distribution of modern glaciers; also, on the example of mountainous weather stations of Georgia, there are considered the climatic conditions (air temperature, atmospheric precipitation, snow cover) that play a decisive role in the maintenance of modern glacial cover.

Keywords

Georgian Caucasus Mountains · Air temperature · Atmospheric precipitation · Snow cover

2.1 Orography and Relief

The high mountainous relief of the Greater Caucasus is favorable for the existence of glaciers in Georgia. The Greater Caucasus mountain range is stretched along the territory of Georgia at ~750 km; the glaciers are concentrated in the southern and partly, in the northern slopes of the watershed range, as well as in the side range and branch ranges of the Greater Caucasus. According to the morphological and morphometric characteristics the Greater Caucasus can be divided into three parts within Georgia—Western, Central, and Eastern.

The Western Caucasus region includes the part, which is located to the west of the Dalari Pass. It has a sublatitudinal direction in Georgia. The relief of its southern slope is characterized by complex orographic structure. The main watershed range is the highest morphological unit here (Fig. 2.1). The Greater Caucasus branch ranges of Gagra, Bzipi, Chkhalta (Abkhazeti), and Kodori are also sharply distinguished morphologically and morphometrically. Endogenic and exogenic relief-creating processes, which have been acting during the entire Neogene–Quaternary period, participated in the formation of modern relief of the Western Caucasus (Geomorphology of Georgia 1971).

The main watershed range. Its morphological characteristics are as follows: the orographic certainty, longitudinal extent, horst-anticline

L. Tielidze, *Glaciers of Georgia*, Geography of the Physical Environment,
DOI 10.1007/978-3-319-50571-8_2

Fig. 2.1 Southern slope of the Western Caucasus main watershed range to the east of the Mount Atsgara (*photo by* R. Gobejishvili)

structure on the surface of which the crystalline core is exposed, and slopes asymmetry. The southern slope is a tectonic step, which runs steeply toward the bottom of the alongside gorges (Astakhov 1973).

By morphological features the main watershed range of the Western Caucasus can be divided into three parts in its turn: western, central, and eastern. The central part is the highest, which is stretched among the Marukhi and Klukhori Passes; height of some of the peaks exceeds 3700 m here.

Within the Western Caucasus the numerous submeridional branch ranges are separated from the main watershed range. Among them can be noted Chibikha, Khita, Khutia, Klichi, and Ghvaghva. They are the watersheds of the basins of the Kodori River right tributaries. The heights of the individual peaks exceed 3000 m.

The axial section of the main watershed range is composed of crystalline rocks of Lower Palaeozoic and Precambrian ages. The mountainous region's line built by these rocks gradually becomes narrow from the east to the west. The following two formations can be distinguished by petrographic composition: (1) crystalline shales, gneisses, phyllites, and marble of the Precambrian and Lower Paleozoic age, into

which sometimes the granite intrusions are intruded; and (2) the Lower Paleozoic granitoids (Geology of the USSR. V. X).

Weathering and exaration action of modern and old glaciations are important in formation of the crest of the main watershed range. The result of the exogenous processes is the presence of frequent comb-like forms with the pointed peaks (carlings), narrow and deep passes, and quite wide glacial cirques. A crest of the main watershed range is located in the nival zone above the firn line and its morphological features create favorable conditions for the formation and existence of modern glaciers there.

The Kodori Range is a well-expressed orographic unit in the Western Caucasus (Fig. 2.2). Height of some of the peaks exceeds 3600 m here. Central part of the range has the latitudinal direction; its height increases and some peaks (M. Khojali) exceed 3000 m. Its western part is rather long and its hypsometric indices are behind the others. The central and western parts are composed by the Bajocian porphyrites. Morphostructurally it belongs to the Kodori and Lechkhumi semi-inversed ranges (Astakhov 1973).

The tectonic and erosion processes have the leading role in the formation of morphosculptural

Fig. 2.2 Kodori range (*photo by* L. Tielidze)

forms of the relief of the Kodori range, and the nival–glacial processes—in the formation of the crest of the range.

The Chkhalta (Abkhazeti) range stretches in parallel to the main watershed range and is 12–17 km away from it. Morphostructurally it is a 7–8-km-wide horst-synclinal range, which has a sublatitudinal direction (Astakhov 1973). The relief is distinguished by a deep fragmentation with the depth of 1000–1500 m. Orographic axis of the range almost coincides the synclinorium axis, which is composed of Bajocian porphyrites and intrusive and effusive rocks of the Mesozoic age; the slopes of the syncline are composed of the Upper Lias shales. The synclinorium is bordered by fault lines on the north and south. As it was mentioned above, the northern fault line is presented as a tectonic step. In the lower mountain zone the erosion–accumulation processes have a leading role in the creation of morphological forms of the Chkhalta range relief, while the upper tiers of the range are created by the nival–glacial processes.

The Bzipi range has a latitudinal direction. It is composed of Mesozoic sediments, while its eastern part is of Bajocian porphyrites. The eastern part of the range is known as the Chedimi range and hypsometrically is the highest one. The Mount Khimsa (3034 m) is its highest peak, to the northern slope of which there are the small glaciers. The traces of old glaciation are well preserved in the high places of the Bzipi range.

The Central Caucasus sector is the highest hypsometrically; it is characterized by a complex geological structure and is very interesting by glacial geomorphological point of view. Its western boundary coincides with the Dalari pass and runs along the Enguri–Kodori Rivers' watershed (Kharikhra range), while its east boundary coincides with the Jvari Pass and then runs along the bottom of the river gorges of Tergi–Bidara–Mtiuleti's Aragvi (Maruashvili 1971).

Orographic structure of the southern slopes of the Central Caucasus is different from the similar slopes of the Eastern and Western Caucasus. It is characterized by a large extent and great depth of fragmentation. Erosion and tectonic processes are important in the formation of modern relief, while the glacial processes—in the high mountain zone.

In terms of the glaciers distribution, the several orographic units can be distinguished in the Central Caucasus: the ranges of Svaneti, Samegrelo, Letchkhumi, Shoda-Kedela, and others.

Two sections—Svaneti and Racha-Dvaleti—can be distinguished in the **main watershed range** according to the geomorphological and geological features.

In the entire Greater Caucasus the **Racha-Svaneti** section (Fig. 2.3) is the highest hypsometrically (4000–5000 m), heights of some of the peaks exceed 5000 m. Its crystal core, which is built of Pre-Paleozoic and Paleozoic formations, is uplifted by a dome-block motion and is naked due to erosion processes. One of the main morphostructural elements of the zone is the main overthrust. The fault length falls by the angle of 35–50°, and therefore, the old crystalline core is overthrusted on the Lias shales on the north. Amplitude of the fault (overthrust) makes several kilometers, and its origination took place in the Oligocene (Gamkrelidze 1966). In this section the main watershed range is presented as a horst-anticline morphostructure (Astakhov 1973). This section in the Caucasus is characterized by deep gorges, steep slopes, and active, powerful action of modern relief-forming processes; it is a major center of modern and old glaciations.

The Dvaleti (east) segment (Fig. 2.4) of the main watershed range is behind the Racha-Svaneti Caucasus hypsometrically. Height of some of the peaks reaches 3800–3900 m. The crest of the main watershed range is coiled, landforms of the relief are characterized by soft contours, and above 3200 m the relief is rocky with comb-like forms due to the activity of weathering processes. In structural terms it belongs to the Shovi–Pasanauri subzone of the Mestia–Tianeti zone. It is composed of the Upper Jurassic and Cretaceous ages flysch and is characterized by a very complex and tense tectonics (Geology of the USSR. V. X 1964).

Many branch ranges come out from the Central Caucasus main watershed range, such as Shdavleri, Tsalgmili, Ghvalda, Kareta, Bodurashi, Mkhvrelieti, Gormaghali, Java, Kharuli, etc. Leading role in the formation of their relief belongs to the nival–glacial processes, especially, in the Late Pleistocene and Holocene.

The Samegrelo range has a sublatitudinal direction. By morphological and morphometric signs it is divided into three parts. Hypsometrically the highest are the western and eastern sections, which mostly consist of porphyrites of Bajocian series; in the western section the outcrops of shales and sandstones of the Jurassic age can be seen under the porphyrites in some places. Shape of the relief is formed by rocky and peaky mountains; the height of some of the peaks exceeds 3000 m. The central part of the range is of 300–400 m lower and is composed of Lower Jurassic age easily decaying rocks, which give the soft contours to the relief. The Samegrelo range was formed in the conditions of the eugeosynclinal inversion of the southern slope of the Greater Caucasus and belongs to the semi-inversed ranges (Astakhov 1973). Modern relief of the Samegrelo range was formed in the

Fig. 2.3 Svaneti Caucasus (*photo by* L. Tielidze)

Fig. 2.4 Dvaleti Caucasus (*photo by* R. Gobejishvili)

Pleistocene by interaction of tectonic and erosion processes, and high mountain zone there is a well-preserved trace of glaciation activity in the Late Pleistocene and Holocene periods.

The Svaneti range is distinguished by the height of the relief, as well as the area and the number of glaciers from the other ranges located in the southern slope of the Greater Caucasus (Fig. 2.5). It is divided into three parts morphologically and morphometrically. The height of the eastern section is somewhat lower than that of the central one; here only the Mount Dadashi reaches 3535 m and there are several glaciers in its slopes. The central section of latitudinal direction, which is the highest and the height of some peaks exceeds 3700–4000 m (Mount Lahili, 4009 m), is located among the Lasili and Leshnuri Passes. Almost all glaciers of the Svaneti range are located here. The western section of the Svaneti range is much lower than the central one; heights of its peaks do not exceed 3300 m and there are no modern glaciers there.

Some scholars (Astakhov 1973) consider the Svaneti range as an individual morphostructure—the Svaneti anticlinal range. It is built mainly of shale—sandstone flysch suite of the Lower Jurassic age. The westernmost part of the range is built of Bajocian porphyrites. Outcrop of the Upper Jurassic period carbonate flysch can be found in the Atkveri Pass. Its central part is built of metamorphic formations of Paleozoic age. Characteristics of the building rocks determine mainly the soft forms of the relief here. In the formation of relief forms an active role belongs to the erosion and glacial processes along with the

tectonic ones. Steep and rocky slopes dominate in the nival zone.

The Lechkhumi range has a submeridional direction from the Mount Pasismta (3779 m) to the Mount Chudkharo (3562 m), and the sublatitudinal direction—from the Mount Chudkharo to the west. Hypsometrically the highest is the central part of the range called Chudkharo–Samertskhle massif. The Letchkhumi range is sharply asymmetric. Its northern slope is shorter than the southern one. The relief is built of Jurassic age flysch suites, but the Sori anticline slopes and the Chudkharo–Kupri syncline core are built of Bajocian porphyrites. In structural–lithological terms, the subzone has many common both with the South Caucasus fold system and Georgian block as well. It belongs to the Kodori and Lechkhumi semi-inversed ranges (Astakhov 1973), which experience hard wash out at the neotectonic stage. Along with the other processes the nival–glacial processes play an active role in the relief.

The Shoda-Kedela range is divided by the Rioni River into two parts; the erosion depth reaches 1500–2000 m here. Its crest is much coiled, which is a result of the reversal river erosion. The range is a horst-synclinal range morphostructurally, which is built of the Upper Jurassic and Lower Cretaceous ages carbonate flysch. The range is accompanied with the young fault lines by sides, resulting in the activation and uplifting of Early Alpine synclinorium. These processes are underway even today. The Shoda-Kedela range is quite high, and some of its peaks exceed 3500 m. The nival–glacial

Fig. 2.5 Svaneti range (*photo by* L. Tielidze)

processes are also remarkable along with the erosion–tectonic processes in the creation of morphosculptural forms of the relief.

The Eastern Caucasus (Fig. 2.6). To it belongs the part of the Greater Caucasus Range, which is located to the east of the Georgian Military Road (the Jvari Pass). Both the southern and northern slopes of the Caucasus range get within the Georgia's boundaries. Its maximum width makes 60 km. The Eastern Caucasus is quite high hypsometrically; its peaks—Kuro, Komito, Shani, Amgha, Tebulosmta, etc.— exceed 4000 m, though, because of the relatively dry climate and morphological features of the relief, the modern glaciers are more poorly represented in the Eastern Caucasus than in the hypsometrically lower Western Caucasus.

The Eastern Caucasus is entirely built of shale-sandstone suites of the Jurassic age and Cretaceous flysch. In some areas there are outlets of intrusions of different ages (Dariali granites, diabases of Chaukhi and Chimgha rocks); tectonic structure is complex, and it is characterized by isoclinal and inverted to the south folds. Alongside the fault lines, which are complicated by overthrusts, faultings and shearings are of great importance in the formation of the relief structures.

By orographic point of view, the crossing gorges and ranges of meridian direction prevail in the Eastern Caucasus. This general picture is violated by the Pirikita range and Tusheti depression, which have the general Caucasian direction. Vertical zoning of the exogenous

morphological complexes can be vividly seen in the relief. The nival–glacial zone is presented discretely and is connected to the separate massifs located above 3300–3500 m. According to the morphological and morphostructural signs, here can be distinguished the main watershed and side ranges with their branches and the Tusheti depression.

The main watershed range has a sublatitudinal direction. The highest peak of the range is the North Chaukhi—3842 m and the heights of the peaks—Roshka and Shaviklde are within 3500 m; and the passes are located at a height of 2300–3200 m. The main watershed range is mainly composed of the Jurassic age schists and sandstones. Against the background of the old glacial relief, the modern nival–glacial landscape is connected with the separate massifs in the form of islands.

The Khokhi range. Only the southern slope of its eastern part gets within the boundary of Georgia (Fig. 2.7). It is the highest massif in the eastern Georgia; height of some of the peaks exceeds 4500 m. The lowest pass is located at 3700 m. The Khokhi range is composed of shales and sandstones of the Lower Jurassic age, which are often destroyed by diabase veins. The young effusive series widely participate in the structure of the radially fragmented Kazbegi massif. Shape of the relief is a result of glacier, old glacial, volcanic and erosion forms, and their interrelations.

The intermountain depressions of synclinal origin are located between the main watershed

Fig. 2.6 Eastern Caucasus (*photo by* R. Gobejishvili)

Fig. 2.7 Khokhi range (*photo by* L. Tielidze)

range and its branches. Among them the remarkable are as follows: Zemo and Kvemo Svaneti, mountainous Racha, Liakhvi, and Truso. The relief is built of the Jurassic age suites. The erosion–accumulation and tectonic processes have the main role in the modeling of the relief. The glaciers of the Late Pleistocene period have a certain role in the creation of morphostructural forms.

The Side range is presented in the Eastern Caucasus in the form of separate massifs, the integrity of which is violated by the erosive action of the river. There are quite high peaks in these massifs—Kamghismaghali (3741 m), Chimghismaghali (3851 m), and Tebulosmta (4493 m). The main orographic units are the watersheds of the meridian direction. The main role in the formation of the relief belongs to the tectonic and erosion processes, which were also active before the neotectonic stage. There are nival–glacial processes only on the crests of the ranges and high massifs.

The side range of latitudinal direction and 42-km long, called Pirikita range, is orographically well expressed in the relief from the Mount Tebulosmta to the Mount Diklosmta. It is the highest range in the Eastern Caucasus, and height of some of the peaks exceeds 4000 m. And the modern glaciers are located namely in the slopes of these peaks. Among other processes the nival–glacial processes have important role in the formation of the relief of the Pirikita range. There are intermountain isocline depressions among the main watershed range and the Side range. Such depressions are located in the headwaters of the Asa, Arghuni, and Andaki Rivers. It should be noted a deeply dissected Tusheti depression of latitudinal direction.

The Late Pleistocene glaciation traces are well preserved in the Greater Caucasus other branch ranges of lower hyphsometry (Bzipi, Racha, Kharuli, Mtiuleti) and in the number of the ranges of the southern Georgian highland. We give their brief geomorphological description when considering the old glaciations.

2.2 Climatic Conditions

Complex orographic peculiarities of the relief of Georgia determine the variety of climatic conditions, particularly the strong fragmentation of the mountain gorge relief is notable, as well as the basic orientation of the mountain ranges, the height alternation, and also the exposure and slope inclination. Variability of climate elements is well reflected in the differentiation of natural processes.

Air temperature is a very necessary element of climate and is a leading factor in the snow-ice cover creation. The Caucasian thermal regime is mainly determined by its geographical location, solar radiation, subsurface feature, atmospheric circulation, and relief. Therefore, the air temperature is characterized by high contrast (Kordzakhia 1967).

January is usually the coldest month in Georgia, but in the high mountain regions

(2700–2800 m) February is often the coldest month. Stable frosty periods at a height of 2000–3000 m last from November to May, and above 3000 m from October to July. The average January temperature is −8 °C at a height of 2000 m and the coldest month is −16 °C at a height of 3600 m (Gobejishvili 1995). The average monthly temperature of the warmest month—August—varies from +14 to +17 °C at about 1500 m of altitude, falling to +7.6 and +3.4 °C, respectively, at 2800 and 3600 m (Gobejishvili 1995). Average multi-annual air temperature ranges from +5.9 °C (Mestia, 1906–2013, 1441 asl) to −5.7 °C (Kazbegi, 1907–2009, 3653 asl) (Tielidze 2016).

Amount of **atmospheric precipitation** has a great impact on the scales of modern glaciation. Interaction of atmospheric circulation processes and local physical–geographical factors determines the distribution of precipitation. It is important to notice the close location of the Black Sea and the barrier direction of the Greater Caucasus mountain range, which protects Georgia from the intrusion of the cold air masses from the north. Average multi-annual precipitation ranges from 400 to 4500 mm in Georgia. In addition, amount of atmospheric precipitation decreases from the west to the east and from the south to the north. Orographic features of individual regions, first of all, the height of the relief, exposition, slope inclination and the direction of the river gorges towards the humid air masses violate this regularity (Kordzakhia 1967).

Southern slopes of the Bzipi, Kodori, Samegrelo, Svaneti, Lechkhumi, and Shoda-Kedela ranges and the high mountain zone of the Greater Caucasus are characterized by abundant precipitation. Amount of precipitation is ∼1600–2300 mm. Intermountain depressions are more characterized by aridity—∼900–1600 mm. In the Eastern Caucasus the amount of precipitation varies from ∼900 mm to ∼1600 mm above ∼2000 m and in the depressions—from ∼700 mm to ∼900 mm.

At the height of ∼1000 m the snow cover melting starts in average in the first half of April, and at the height of ∼2000 m—in the middle of May—the snow occurs early in the glaciers surface and melts late. Snow is redistributed under the impact of snow avalanches and wind. Two types of snow redistribution are mainly distinguished: (1) when the snow redistribution occurs within the same basin, and (2) when the snow is redistributed from one to another basin. Existence of the glaciers in the Greater Caucasus is mainly stipulated by the accumulation of excess snow.

Wind. The Caucasus mountain system is characterized by the diversity of wind directions. Western winds dominate throughout the year in the mountain slopes. Wind direction in the crest of the Greater Caucasus is stipulated by the general circulation of the atmosphere; here all year round western moist winds are blown. The seasonality in wind speed direction is well expressed in the high mountain and medium mountain areas.

The average annual wind speed varies from ∼0.4 m/s (Shovi) to ∼6.3 m/s (Kazbegi). The lowest wind speed (∼0.4–2.0 m/s) is in the Greater Caucasus river gorges (Kordzakhia 1967).

Maximum monthly average wind speed in the high mountain areas of the Greater Caucasus is during the cold spell (November–March). The lowest wind speed values are in the summer months (June–August), as there is no mountain gorge circulation in the upper layers. In the river gorges the average monthly wind speed does not exceed ∼2.7 m/s (Ambrolauri), while on the crests of the ranges and passes the wind speed reaches ∼7.6 m/s (Kazbegi) (Kordzakhia 1967).

Strong winds are rare in the deep gorges of the Greater Caucasus Rivers, but they are of one direction, which is stipulated by the layout of the river gorges. The western winds with the speed more than ∼40 m/s dominate in the high peaks and crests. This pattern is particularly well observed during the winter period.

The snow cover is unevenly distributed in the Greater Caucasus range; snowfall in the western Caucasus is greater and snow cover lasts for a longer period than in the eastern Caucasus. The average snow cover in the eastern Caucasus is ∼110 days at ∼1500 m a.s.l., ∼145 days at ∼2000 m a.s.l., and ∼195 days at ∼2500 m a.s.l. In the western Caucasus snow cover extends for ~135, 182, and 222 days, respectively. In the eastern Caucasus the average depth of snow cover is ~21–40 cm at ~1500–2000 m elevation, and more than 100 cm at ~2000–2500 m (Gobejishvili 1995; Tielidze 2016).

Thus, we conclude that the occurrence of the glaciers of Georgia is stipulated by the climatic conditions of its territory, the impact of the Black Sea, geological structure, and the relief's morphometric and morphological peculiarities.

References

Astakhov NE (1973) Structural geomorphology of Georgia. Pub. House "Metsniereba". Tbilisi (in Russian)

Gamkrelidze PD (1966) The main features of the tectonic structure of the Caucasus. Geotectonics 8 (in Russian)

Geology of the USSR. V. X (1964) Georgian SSR. Part I. "Nedra". Moscow (in Russian)

Geomorphology of Georgia (1971) Pub. House "Metsniereba". Tbilisi (in Russian)

Gobejishvili RG (1995) The evolution of the modern ice age glaciers and mountains of Eurasia in the Late Pleistocene and Holocene. The thesis of doctor of science degree in geography (in Georgian)

Kordzakhia R (1967) Enguri and Tskhenistskhali River basins climate features within Svaneti. Acts Georgian Geogr Soc IX–X:110–125 (in Georgian)

Maruashvili LI (1971) Physical geography of Georgia, Monograph. Publ. House "Metsniereba". Tbilisi (in Georgian)

Tielidze LG (2016) Glacier change over the last century, Caucasus Mountains, Georgia, observed from old topographical maps, Landsat and ASTER satellite imagery. Cryosphere 10:713–725. doi:10.5194/tc-10-713-2016

Abstract

This chapter discusses the glaciers' new inventory methodology. Digital-ization methods are developed using both the old topographic maps and modern satellite images. The error of identification of the areas of glaciers is analyzed. All glacier basins and large glaciers are described in details as well.

Keywords

Glacier inventory · Remote sensing · Georgian Caucasus glaciers

3.1 New Inventory Data Sources and Methods

3.1.1 Old Topographical Maps

The compilation of the first reliable map of the Caucasus, at a scale of 1:420,000 and depicting the largest glaciers, was completed by 1862. Topographic surveys of the Caucasus at a scale of 1:42,000 were accomplished 50-years later (1880–1910). Having analyzed these maps, Podozerskiy (1911) published the first inventory of Caucasus glaciers (Kotlyakov et al. 2010). Detailed analysis of these early data showed some defects in the shape of the glaciers and in particular the inaccessible valley glaciers were depicted incorrectly. This caused some error in the identification of precise areas, such as in the Enguri and Rioni River basins, which were difficult to access for plane table surveying. However as no other data exist from this time, these maps are the most reliable source for this research to establish century-long trend glacier changes (Tielidze 2016a).

The oldest topographic maps were replaced in 1960, under the former Soviet Union with 1:50,000 scale topographical maps from 1955 to 1960 aerial images. Based on these, Gobejishvili (1989) generated new statistical information on the glaciers of Georgia.

The next inventory of the Caucasus glaciers was the result of a manual evaluation of selected glacier parameters from the original aerial pho-tographs and topographic maps (Catalog of Glaciers of the USSR 1975; Khromova et al. 2014), where information on Georgia was obtained from the same (1955–1957) aerial photographs. There are some mistakes in the

L. Tielidze, *Glaciers of Georgia*, Geography of the Physical Environment,
DOI 10.1007/978-3-319-50571-8_3

catalog regarding number and area of the glaciers in some river basins (particularly the Bzipi, Kelasuri, Khobisckali, Liakhvi, Aragvi, and Tergi), where temporary snowfields were considered as glaciers (Gobejishvili 1995). The USSR catalog data sets contain tables with glacier parameters but not glacier outlines.

As this information was only available in printed form, we scanned and coregistered the maps and images using the 3 August 2014 Landsat image as a master. Offsets between the images and the archival maps were within one pixel (15 m) based on an analysis of common features identifiable in each data set. We reprojected both maps (1911, 1960) to Universal Transverse Mercator (UTM), zone 38-north on the WGS84 ellipsoid, to facilitate comparison with modern image data sets (ArcGIS 10.2.1 software).

3.1.2 Landsat and ASTER Imagery and Glacier Area Mapping

Many of the world's glaciers are in remote areas, such that land-based methods of measuring their changes are expensive and labor-intensive. Remote sensing technologies have offered a solution to this problem (Kaab 2002). Satellite imagery-Landsat L8 OLI (Operational Land Imager), since February 2013, and Advanced Spaceborne Thermal Emission and Reflection Radiometer (ASTER), since January 2000, with 15/30 m resolution provide convenient tools for glacier analysis. Together with old topographical maps, these allow us to identify changes in the number and area of glaciers over the last century. Most of the images (Landsat and ASTER) were acquired at the end of the ablation season, from 2 August to 2 September (except for one ASTER image, on 10 July), when glacier tongues were free of seasonal snow under cloud-free conditions and suited for glacier mapping (Fig. 1.2), but with some glacier margins obscured by shadows from rock faces and glacier cirque walls (Khromova et al. 2014). Landsat (level L1T) georeferenced images were supplied by the US Geological

Survey's Earth Resources Observation and Science (EROS) Center and downloaded using the EarthExplorer tool (http://earthexplorer.usgs.gov/). ASTER (level L1T) images were supplied by the National Aeronautic and Space Administration's (NASA) Earth Observing System Data and Information System (EOSDIS) and downloaded using the Reverb/ECHO tool (http://reverb.echo.nasa.gov/).

Following the Tielidze (2016a) we used the Landsat 8 panchromatic band, along with a color-composite scene for each acquisition date, for Landsat images—bands 7 (short-wave infrared), 5 (near-infrared), and 3 (green); for ASTER images—bands 3 (near-infrared), 2 (red), and 1 (green). Each glacier boundary was manually digitized and the total surface area for each glacier calculated according to Paul et al. (2009). The size of the smallest glacier mapped was 0.01 km^2.

3.1.3 Glacier Delineation Error and Analysis

For the Georgian Caucasus glaciers we calculated three error terms resulting from (a) coregistration of old maps and satellite images, (b) glacier area error and (c) debris cover assessment.

(a) Offsets between the images and archival maps are within 1 image pixel (15 m). Glacier outlines on the old topographic maps (1911, 1960) correspond to a line thickness of 12 m (1:42,000) and 15 m (1:50,000). Using the buffer method from Granshaw and Fountain (2006), these yields a total potential error of ±1.64%.

(b) The glacier area error is mostly inversely proportional to the length of the glacier margin (Pfeffer et al. 2014). Applying glacier buffers account for the length of the glacier perimeter, while the buffer width, is critical to the resultant glacier area error (Guo et al. 2015). We estimated uncertainty by the buffer method suggested by Bolch et al. (2010) and Granshaw and Fountain

(2006) with a buffer size 7.5 m for all aerial images and maps, based on the 15 m image pixel size, and map uncertainty in the absence of stated historical accuracies. This generated an average uncertainty of the mapped glacier area of 2.3% for 2014 (satellite images), 2.0% for 1960 (topographical maps) and 1.6% for 1911 (Podozerskiy catalog).

(c) Manual digitizing by an experienced analyst is usually more accurate than automated methods for glaciers with debris cover (Raup et al. 2007), which is a major source of error in glacier mapping (Bhambri et al. 2011; Bolch et al. 2008) In the Caucasus, supra-glacial debris cover has a smaller extent than in many glacierized regions, especially Asia (Stokes et al. 2007; Shahgedanova et al. 2014). One of the most heavily debris-covered glaciers in the Georgian Caucasus is Khalde Glacier (42.596°N, 43.22°E) where supra-glacial debris covers 23%. For the precise determination of debris cover, we also used our GPS field data collected in most glaciated areas during 2004–2014, including those with highest debris cover (Khalde, Lekhziri, Chalaati, Shkhara, Devdoraki, Zopkhito, Ushba, Buba, Gergeti,

and Abano). Thus, the error associated with debris-covered glaciers was considered to be negligible.

3.2 General Description of the Modern Glaciers

The spatial distribution of the modern glaciers in the territory of Georgia is stipulated by the peculiarities of atmospheric processes, morphological-morphometric conditions of the relief and their interaction. Main centers of glaciation are related to the elevated Greater Caucasus watershed range and Kazbegi massif. Individual centers can be found in the Greater Caucasus branch ranges of Bzipi, Kodori, Samegrelo, Svaneti, Lechkhumi, Pirikita, etc.

According to the data of 2014, there are 637 glaciers in Georgia with a total area of 355.80 ± 8.25 km^2 (Tielidze 2016a). Modern glaciers are mainly concentrated in the Enguri, Rioni, Kodori, and Tergi River basins, where there are the peaks of 4500 m and higher. 89.32% of the total amount and 97.15% of the total area of glaciers of Georgia are located in these basins (Table 3.1).

Table 3.1 Distribution of the glaciers of Georgia according to the river basins

Basin name	Landsat and ASTER Imagery, 2014		
	Number	Area, km^2	Uncertainty (%)
Bzipi	18	3.99 ± 0.13	± 3.25
Kelasuri	1	0.11 ± 0.005	± 4.54
Kodori	145	40.06 ± 1.29	± 3.22
Enguri	269	223.39 ± 4.6	± 2.05
Khobisckali	9	0.46 ± 0.03	± 6.52
Rioni	97	46.65 ± 1.15	± 2.47
Liakhvi	10	1.82 ± 0.07	± 3.84
Aragvi	1	0.31 ± 0.015	± 4.83
Tergi	58	35.56 ± 0.8	± 2.24
Asa	3	0.54 ± 0.025	± 4.62
Arghuni	6	0.43 ± 0.025	± 5.81
Pirikita Alazani	20	2.42 ± 0.11	± 4.54
Total	637	355.80 ± 8.25	± 2.32

Modern glaciers are unevenly distributed between the different river basins. The leading place belongs to the Enguri River basin; 42.22% of the total number of the glaciers of Georgia is the share of it, as well as 62.78% of the total area of the glaciers of Georgia is a share of the Enguri River basin.

Except the Enguri River basin the shares of the other river basins in the total number of the glaciers of Georgia are distributed as follows: the Kodori River basin—22.76%; the Rioni River basin—15.22%; the Tergi River basin—9.10%; the Pirikita Alazani River basin—3.19%; the Bzipi River basin—2.82%; the Liakhvi River basin—1.56%, and the Khobistskali River basin—1.41%. As for the basins of the rivers of Arghuni, Asa, Aragvi, and Kelasuri, their joint share is 1.72% in the total number of the glaciers of Georgia.

Except the Enguri River basin, the shares of the other river basins in the total area of the glaciers of Georgia are distributed as follows: the Rioni River basin—13.11%; the Kodori River basin—11.25%; the Tergi River basin—9.99%; the Bzipi River basin—1.12% and the Pirikita Alazani River basin—0.68%. As for the basins of the rivers of Liakhvi, Asa, Khobistskali, Arghuni, Aragvi, and Kelasuri, their joint share is 1.07% in the total area of the glaciers of Georgia.

In Georgia, the glaciers are unevenly distributed, both as orographical units and according to the individual river basins. We will consider them according to the river basins in order to have a clearer picture of the glaciers of Georgia.

3.2.1 Glaciers of the Bzipi River Basin

The Bzipi river gorge is the westernmost basin of the territory of Georgia, where there are the modern glaciers. The basin is located between the Western Caucasus and Bzipi ranges. The modern glaciers are located in the southern slope of the Greater Caucasus and the northern slope of the Bzipi range. Although the relief is not hypsometrically high (it is represented by about 3400 m high peaks), the distribution of the modern glaciers is stipulated by their location near the Black Sea, circulation of air masses and morphological and geological conditions of the relief. In the Bzipi River gorge the firn line is located very low (3000 m).

According to Podozerskiy (1911), there were 10 glaciers with the total area of 4.03 ± 0.085 km^2. But according to the topographic maps of 1946–1950, there were 18 glaciers with the total area of 9.36 km^2. According to the 1960 data, there were still 18 glaciers with the area of 9.90 ± 0.20 km^2. According to the 2014 data, the number of glaciers are the same but the area was reduced to 3.99 ± 0.13 km^2. It should be noted that the Bzipi River basin is the only one in Georgia, where the number of glaciers has not been changed since 1946 (with the exception of the Kelasuri River basin, where today there is still one glacier like it was in 1960). But if we judge by the data of Podozerskiy, it follows that during the last century the number of the glaciers in the Bzipi River basin has increased and the area remained the same, that we think is illogical, as there were made certain mistakes in some cases in the defining the number and area of glaciers during the map compilation by Podozerskiy. And, the fact that the amount of glaciers has not been changed during the 68 years, it seems, is connected with the relief and climatic conditions necessary for the existence of glaciers. But, the global climate change has been vividly observed here as well, especially in recent decades, as the area of the glaciers was reduced by $\sim 57.27\%$ in 1946–2014.

The Bzipi River basin is characterized by the small sizes of the glaciers: there are 16 glaciers with the area of ~ 0.5 km^2, total area of which is 1.80 km^2. The other two glaciers are of cirque-valley types, one of which is located on the southern slope of the Greater Caucasus, and the other—in the northern slope of the Bzipi range. Their total area is 2.20 km^2.

According to exposition, the glaciers of all expositions are represented in the Bzipi River basin except of the northeastern one. Their distribution is given in Fig. 3.1.

Fig. 3.1 Distribution of the glaciers in the Bzipi River basin by exposition

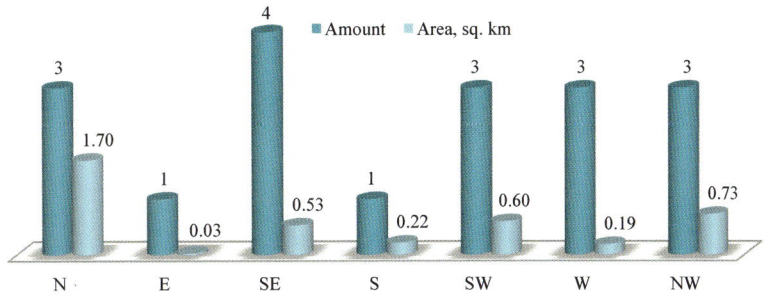

3.2.2 Glaciers of the Kelasuri River Basin

The Kelasuri River basin originates from the southern slope of the Bzipi range. By the data of 2014, there is one glacier with an area of 0.11 ± 0.005 km^2. K. Podozerskiy does not provide any kind of information about the glaciers in this basin. By the data of 1960, the area of the same glacier was 0.26 ± 0.015 km^2. Its morphological signs indicate that this glacier has changed considerably during the last stade glaciation. Morphologically, the glacier is of cirque type with the northeastern exposition.

3.2.3 Glaciers of the Kodori River Basin

The main center of glaciation in the Kodori River basin is the southern slope of the Western Caucasus—from the Marukhi Pass to the Dalari Pass. The high peaks, such as: Marukhi, Ertsakho, Sopruju, Dombai Ulgeni, Khakeli, and Ghvandra, are located in this section. Their heights exceed 3800–4000 m. We have the isolated centers of the glaciations in the Kodori, Chkhalta (Abkhazeti), Khutia, and Klichi ranges. Especially, remarkable is the western and south-western slopes of the Kharikhra range, which is a main center of the glaciation of the Sakeni River basin (Fig. 3.2). The peaks higher than 3700 m are located in this mountain range, such as Kharikhra, Maguashikhra, and Okrilatavi (Tielidze et al. 2015c).

Study of the glaciers of the Kodori River basin began even in the last century. Information about

glaciers give us the famous researchers of the Caucasus: Radde (1873), Podozerskiy (1911), Bush (1914), Ivankov (1959) and Tabidze (1965).

A lot of work was conducted by D. Tabidze to study the glaciers of the research region; he studied the glaciers of the Kodori River basin and gave us a new catalog of the glaciers, which is mainly based on the 1:50,000 scale topographic maps drawn up in 1952 (although it should be indicated, that the glaciers are very poorly depicted in the mentioned maps; the maps are drawn up based on aerial photographs of 1946).

According to the data of K. Podozerskiy, there were 118 glaciers in the Kodori River basin with the total area of 73.20 ± 1.55 km^2; according to the data of D. Tabidze—there were 141 glaciers with the total area of 60.0 km^2, and by the data of topographic maps of 1960—there were 160 glaciers with the total area of 63.73 ± 1.63 km^2. By the data of 2014, there are 145 glaciers in the basin with a total area of 40.06 ± 1.29 km^2. The glaciers are unevenly distributed in the Kodori River basin, not only according to the orographical units, but also according to the individual tributary river basins.

Here dominate small glaciers.[1] The glaciers are distributed according to the tributary river basins (Fig. 3.3).

The glaciers of all morphological types except the complex-valley type are distributed in the Kodori River basin. Morphological and morphometric characteristics of the relief and climatic conditions are favorable for existence of

---

[1]The glaciers of Georgia can be divided into three groups according to the area: (1) Small glaciers by the size of ~ 0.5 km^2; (2) Average glaciers—from 0.5 to 2.0 km^2; (3). Large glaciers—over 2.0 km^2.

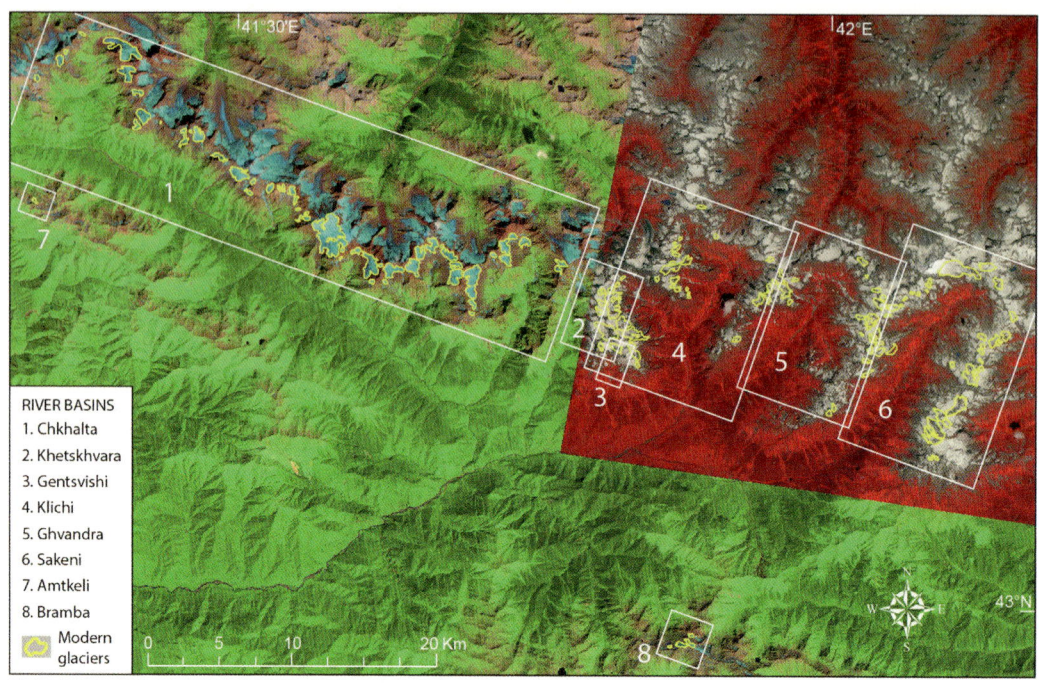

Fig. 3.2 Distribution of modern glaciers of the Kodori River basin according to the tributary river basins

Fig. 3.3 Distribution of the glaciers of the Kodori River basin according to the tributary river basins

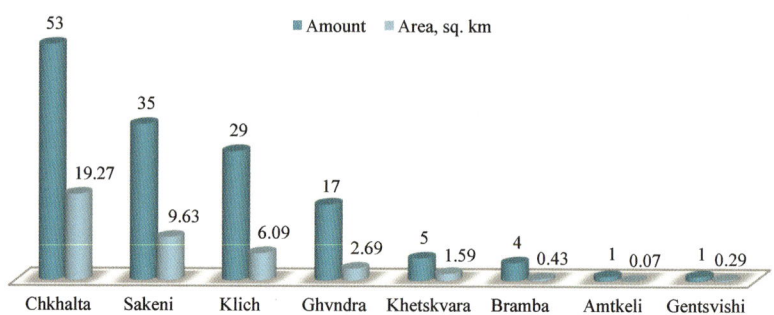

the multiple cirque type glaciers. The cirque type glaciers occupy the first place according to the number and area. 66.20% of the total number and 48.10% of the total area of the glaciers of the Kodori River basin is a share of the cirque type glaciers (Fig. 3.4).

According to the area, the second place occupy the glaciers of cirque-valley type and according to the number, the second place occupy the cirque-hanging glaciers.

There are the glaciers of all expositions in the Kodori River basin (Fig. 3.5). By number exceed the glaciers of northwestern exposition—

31.03%, then comes the glaciers of the eastern exposition—15.17% and south-eastern expositions—14.48%. According to the occupied area, there is a same sequence: share of the glaciers of northwestern exposition in the Kodori River basin exceeds 30.47%, eastern exposition—19.64% and south-eastern—14.12%.

The Kodori River basin is located on the southern slope of the Greater Caucasus and the exposition should be mainly of southern direction, but the conducted study shows that the glaciers of commonly northern exposition dominate, both in number and area. Predominant

Fig. 3.4 Distribution of the glaciers in the Kodori River basin according to the morphological types

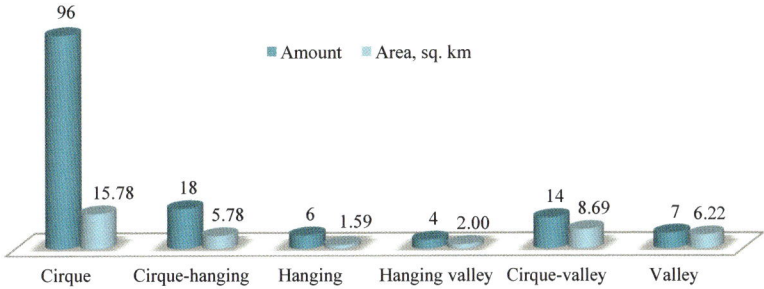

Fig. 3.5 Distribution of the glaciers in the Kodori River basin according to the exposition

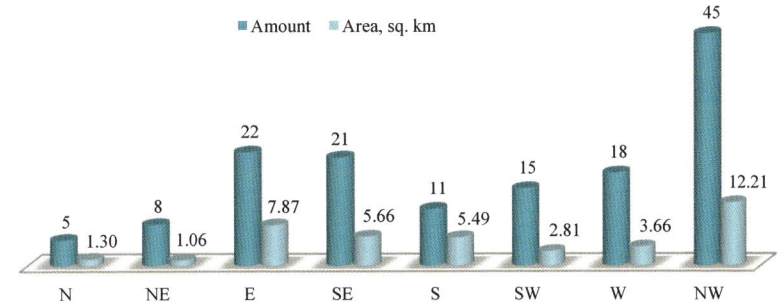

distribution of the glaciers of the northern exposition can be explained by the sublatitudinal or latitudinal direction of the branch ranges.

As it was noted above, the glaciers of the Kodori River basin are not equally distributed, not only according to the orographical units, but also according to the individual river basins; therefore, to have a clearer idea on the modern glaciers of the Kodori River and its tributary river basins and their changes in time and space, it is necessary to consider these basins separately.

The Chkhalta River basin. 36.55% of the amount and 48.10% of the area of the glaciers of the Kodori River basin are the shares of the Chkhalta River Basin. Similarly as in the entire Kodori basin, here the glaciers are distributed unequally by the morphological types and exposition (Figs. 3.6 and 3.7). Two centers of the glaciation can be distinguished in the Chkhalta River basin.

The Marukhi River basin. A separate description has its preconditions: (1) It is a completely separate center in terms of the modern glaciation. (2) According to the old glaciations this basin was independent as well.

According to K. Podozerskiy, there were 12 glaciers in the Marukhi River basin with the area of 6.36 km^2. By the data of 1960, there were 10 glaciers in this basin with the area of 4.82 km^2. As for the data of 2014, there are nine glaciers with the total area of 3.85 km^2.

Basin's largest glacier is Marukhi, which belongs to the type of valley glacier. Its area is 1.68 km^2 and the length is 3.0 km. The Marukhi glacier has a clearly defined ice tongue and 300 m high icefall. The ice tongue is of ∼1.0 km long, which is weakly fractured and characterized by a slight inclination. The central part of the tongue is clean, only its last section is covered with loose material. The surface of the ice tongue is rugged by melting waters due to its weak fracturing and slight inclination. Water-raised grooves are quite large (2–5 cm wide) and they can be found only on the pure glacier surface. In front of the ice tongue at a distance of ∼800–900 m long, in the valley there are moraine hillocks of different sizes, which allow us to rebuild the location of the glacier during the LIA maximum and discuss its dynamics for the last ∼200 years.

Fig. 3.6 Distribution of the glaciers in the Chkhalta River basin according to the morphological types

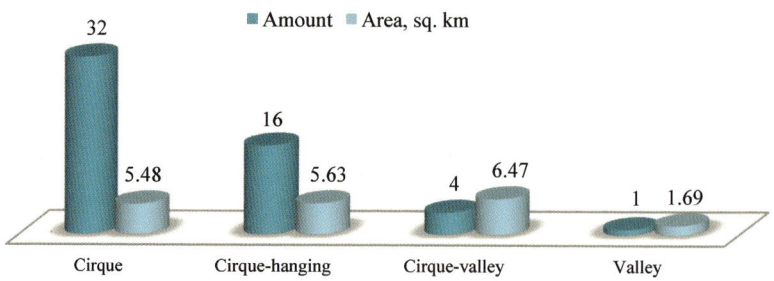

Fig. 3.7 Distribution of the glaciers in the Chkhalta River basin according to the exposition

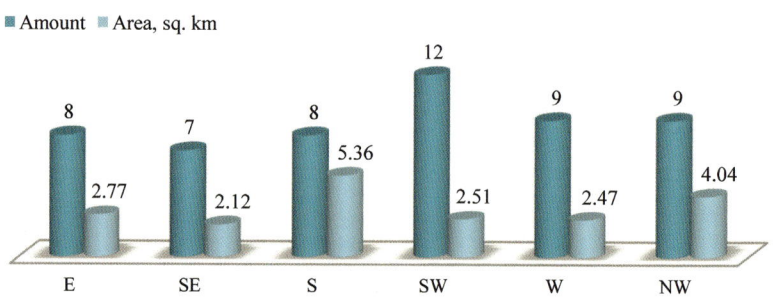

It is true that the Marukhi glacier has no lateral stade moraines, but the trace of the stade glacier is distinguished on the both slopes of the valley by a sharp line (according to the moraine material spread on the slope, vegetation cover, slope color tone, and other morphological parameters). The last stade and micro-stade moraines are well expressed in the valley. The last LIA stade moraine covers the top of the ledge and is presented by ∼10–15 m thick loose material. Thickness of the micro-stade moraines is ∼1.3 m. There is a ∼150–200 m long straight surface between the micro-stade moraines, which is built by the fluvioglacial deposits. As a whole, there are three micro-stade moraines in the valley with two rows per each.

The rest of the glaciers of the Marukhi River Basin are of small cirque and cirque-hanging types. They are mainly located in the headwaters of the rivers of Buloni and Chvakhra. These basins fed the large Marukhi glacier during the Wurm glaciation together with the Marukhi River basin.

The left section of the Chkhalta River gorge from the Mount Ertsakho to Mount Ptishi covers the southern slope of the Greater Caucasus. The peaks create good conditions for existence of glaciers. The glaciers are of cirque,

cirque-hanging, and cirque-valley types. According to the data of K. Podozerskiy (1911) there were 25 glaciers there with the total area of 19.1 km^2. By the data of 1960, there were 32 glaciers with the total area of 15.56 km^2. By the data of 2014, there are 25 glaciers with the total area of 8.10 km^2. The comparison between the mentioned data well depicts the degradation of glaciers in recent decades.

In the given section, the biggest glacier is the **Southern Sopruju** (Fig. 3.8). The glacier has an extensive firn field, its ice tongue is characterized by small parameters and it is hanged over the ledge. The glacier is slightly fractured, let alone the last section of the ice tongue. Due to the morphological conditions of the area, there is a lack of the moraine material in the glacier (both the bed moraine and surface moraine material). The glacier exarates the bedrocks. The Sopruju glacier is the largest in size in the whole Kodori River basin; its area is 3.66 km^2.

The Atsiash-Atsgara River headwater is distinguished by the independent center of glaciation, which morphologically represents the old cirque. By 2014, data here are the eight glaciers. All eight glaciers are of cirque-hanging type. Their total area is 3.75 km^2.

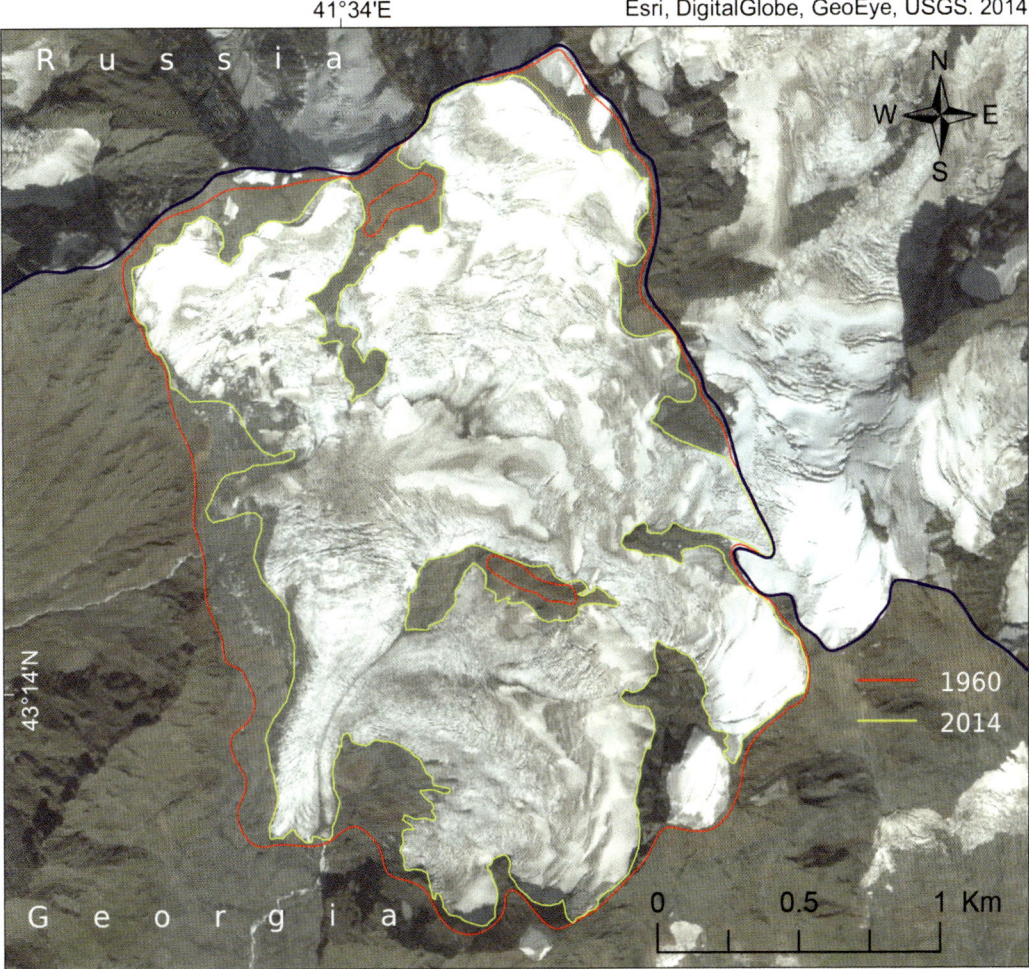

Fig. 3.8 Sopruju glacier reduction in 1960–2014

The Ptishi River is a left tributary of the Chkhalta River. The morphology and morphometric features of the relief of its headwaters create the favorable conditions for the existence of glaciers. By the data of 1960, there were 12 glaciers with the total area of 4.0 km^2, and by the data of 2014 there are seven glaciers with the total area of 1.78 km^2.

The glaciers in the **Khetskvara River basin** are developed on the western slope of the Khutia range. By K. Podozerskiy there were six glaciers in this basin with the total area of 2.8 km^2. By the data of 1960, the number of the glaciers and their area remained unchanged. By the data of 2014, there are five glaciers with the total area of

1.58 km^2. Four glaciers are of cirque type morphologically, and one is of hanging type.

Due to the fact that the glaciers are distributed on the western slope of the Khutia range, their exposition is of northwestern and western.

According to the topographical maps of 1960, there were two small cirque type glaciers of the southern exposition in the **Gentsvishi River basin**. There is no information on them in the work of K. Podozerskiy. By the data of 2014, there is a small cirque type glacier of southeastern exposition.

The **Klichi River basin** is the only gorge in the Kodori River basin, which maintains the shape of the trough valley along the whole distance. The modern glaciers descent to a very low elevation. The glaciers are mainly represented in the basins of the rivers of Achapara, Klichi, and Nakhari.

According to K. Podozerskiy, there were 25 glaciers with the area of 16.55 km^2 in the Klichi River basin at the end of the nineteenth century. By the data of 1960, the number of the glaciers was 28 with the total area of 9.45 km^2. We should indicate that the distribution of the glaciers in the topographical map of 1960 and of K. Podozerskiy do not coincide to each other. The glaciers presented in the Catalog of K. Podozerskiy were not reflected in the data of 1960. It is related to the small glaciers or the snow spots, which have disappeared due to the intense retreat of the glaciers or they still are the snow spots (in the Achapara gorge and near the Klukhori Pass). By the data of 2014, the number of glaciers is 29 with the total area of 6.09 km^2. The total area of the glaciers has reduced by 35.56% after the year of 1960 and their number has increased by one.

There are mainly small cirque type glaciers in the Klichi River basin. The glaciers located in the Khutia range are of eastern exposition; as for the glaciers located on the Klichi range, their exposition is of western or northwestern direction. The glaciers are distributed by morphological types and exposition as follows (Figs. 3.9 and 3.10).

20.0% of the total number and 15.20% of the total area of the glaciers of the Kodori River basin is a share of the glaciers of the Klichi River basin. It is behind of the basins of the rivers of Chkhalta and Sakeni by the thickness of the ice sheet. In the Klichi River basin the largest is the

Klichi glacier. Its area is 0.76 km^2. The glacier starts from the Mount Klichi and spreads to the northeastern direction. Its ice tongue ends at a height of 2450 m in the shape of a forehead; and in front of it there are well-expressed annual moraine hillocks, the heights of which do not exceed ∼0.5–1.0 m. These forms are created due to every year movement of the glaciers and they fix the size of the glaciers retreat in time. The number of the arc wise hillocks is 15 and they end at the top of the ledge, but the annual moraines below are washed away and the bedrocks are outcropped. The moraines are located in parallel to each other and sometimes they cover each other. Distance among the annual moraines is measured (Gobejishvili 1995) in the different places and their distribution from the ice tongue is as follows (Table 3.2).

Lateral stade moraine, very deformed due to the impact of snow avalanches, stretches along the left side of the Klichi glacier. This moraine passes into the relief, also into the last, badly remained moraine. There is a vertical wall to the right of the glacier and the lateral moraine hillocks are not developed.

There are three cirque type glaciers to the left of the Klichi glacier. The glaciers exarate the bedrocks. Their bottom is quite inclined and the glaciers are not able to develop the moraines. The glaciers end nearly at a similar elevations.

The cirque type glaciers have retreated by ∼400–500 m after the LIA maximum, on which indicate the well-developed lateral moraine hillocks in front of the glaciers and the lateral moraines, which are extremely washed away. Among these three glaciers the right-hand glacier

Fig. 3.9 Distribution of the glaciers in the Klichi River basin according to the morphological types

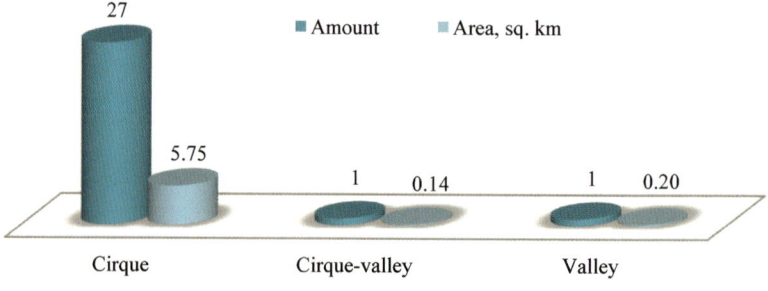

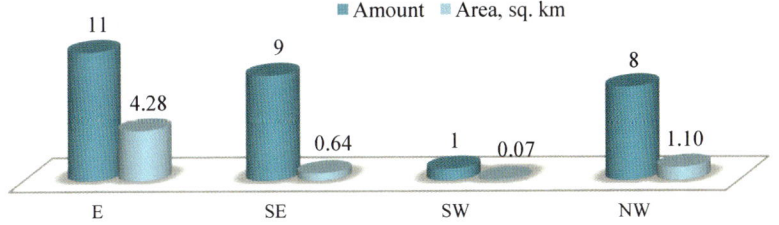

Fig. 3.10 Distribution of the glaciers in the Klichi River basin according to the exposition

Table 3.2 Distribution of the annual moraines of the Klichi glacier

Amount of moraines	Distance from the ice tongue to the moraine	Distance among the moraines	Time of origination (year)
From the tongue—I	4.10	4.10	1975
I–II	6.0	1.90	1974
II–III	9.80	3.80	1973
III–IV	14.0	4.20	1972
IV–V	16.5	2.5	1971
V–VI	17.4	0.9	1970
VI–VII	18.4	1.0	1969
VII–VIII	20.0	1.6	1968
VIII–IX	21.0	1.4	1967
IX–X	22.5	1.1	1966
X–XI	24.0	1.5	1965
XI–XII	26.0	2.0	1964
XII–XIII	28.0	2.0	1963
XIII–XIV	30.0	2.0	1962
XIV–XV	32.3	2.3	1961
Total		32.3	15 years

has very weakly developed moraines. It was a Klichi glacier branch in the LIA maximum. Its bottom is inclined and does not allow the formation of moraines.

All early researchers mention the two glaciers to the right of the Klukhori Pass. Today there are powerful snow covers instead of them. Morphostructural character of the relief indicates that in the past (nineteenth century) here, for sure, should be the small glaciers of a cirque type, which are now melted and there are thick snow covers instead of them.

In the **Ghvandra River basin** there were 24 glaciers with the total area of 7.20 km^2 by K. Podozerskiy; by the data of D. Tabidze—20 glaciers with the total area of 6.0 km^2, and by the data of 1960–22 glaciers with the total area of 6.20 km^2. By the data of 2014, there are 17 glaciers in this basin with a total area of 2.69 km^2.

The Ghvandra River gorge has a shape of typical trough above the site of Mindora. Nevertheless, the Late Holocene stade moraines are poorly remained due to the active erosion and gravitation processes.

The glaciers are distributed by morphological types and expositions as follows (Figs. 3.11 and 3.12).

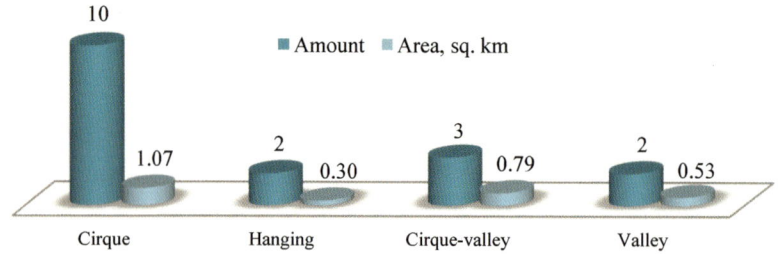

Fig. 3.11 Distribution of the glaciers in the Ghvandra River basin according to the morphological types

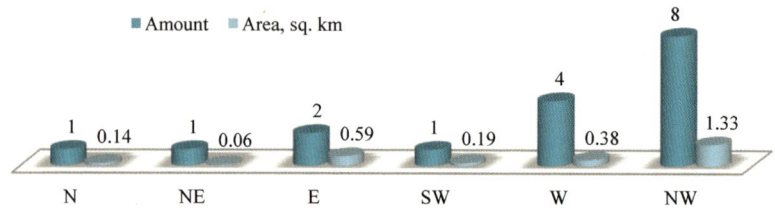

Fig. 3.12 Distribution of the glaciers in the Ghvandra River basin according to the exposition

The Sakeni River basin is located between the ranges of the Greater Caucasus and Kodori. Its area is 233 km². Glaciers are mainly located in the southern slope of the higher elevated Greater Caucasus and the western slope of the Kharikhra range. There are small glaciers in the northeastern slope of the Ghvaghva range. Sakeni River gorge is a second by the number and area of the glaciers among the Kodori River tributaries. 24.13% of the number and 24.03% of the area of all glaciers of the Kodori basin is a share of the Sakeni River basin.

By K. Podozerskiy, there were 17 glaciers in the Sakeni River gorge with the total area of 15.51 km²; by the data of 1960 there were 36 glaciers with the total area of 19.1 km², and by the data of 2014 there are 35 glaciers with the total area of 9.63 km².

Increase in number of the glaciers in 1911–1960 was stipulated by their division during a degradation of ice cover and by their complete depiction on the topographical maps; and the increase in area of the glaciers was related with the indication of reduced areas in the maps of the times of K. Podozerskiy. By the data of 1946–1950, the area of the glaciers was 27.18 km² and their number was 30. Analysis of these two

periods (1946 and 1960) shows that the number of the glaciers has increased and the areas have reduced. These data clearly reflect the situation, which was typical for the evolution of the glaciation of the Greater Caucasus before the years of 1960–1980, when in parallel with the reduction in area of the glaciers the increase in their number took place. We cannot say it today, because, as we mentioned above, for about last half a century the decrease in number of glaciers takes place in the Greater Caucasus together with the reduction in their area. The same situation is in the Sakeni River basin, when in 1960–2014 the area of the glaciers has reduced by 49.59%, while the number of the glaciers has decreased by one.

As for the morphological types, here can be found all types of glaciers except of the valley complex type. They are distributed as follows (Fig. 3.13).

Glaciers exposition is mainly northern (N, NE, NW). This is due to the abundance of glaciers on the western slopes of the Kharikhra mountain range (Fig. 3.14).

The Sakeni glacier, which is located in the southern slope of the Greater Caucasus, is the largest by area in the Sakeni River basin. It is a valley type glacier and has the south-eastern

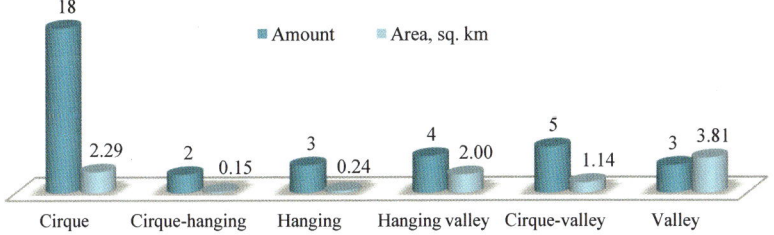

Fig. 3.13 Distribution of the glaciers in the Sakeni River basin according to the morphological types

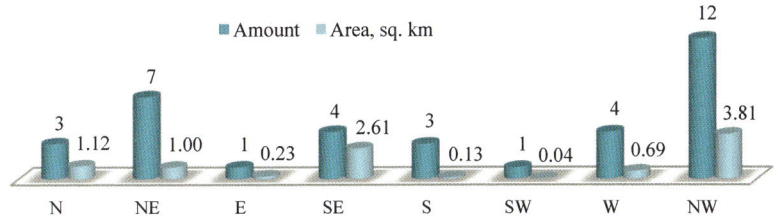

Fig. 3.14 Distribution of the glaciers in the Sakeni River basin according to the exposition

exposition. In 1960, the glacier area was 2.50 km^2 and by the data of 2014 its area is 1.96 km^2. The glacier area was reduced by of 21.60% in the years of 1960–2014 (Fig. 3.15).

The Brakhma River basin is located on the northern slope of the central section of the Kodori range. The range has a latitudinal direction and is not high hypsometrically (Mount Khojali—3314 m), which causes a lack of glaciers. There are four glaciers in this basin with the total area of 0.43 km^2. Three glaciers are of cirque type, and one—of cirque-valley type. Glaciers are distinguished with their small sizes. Their expositions are of northern and northwestern direction. K. Podozerskiy did not mention the glaciers in this basin. Though there were four glaciers in the Brakhma River basin by the data of 1960 with a total area of 1.16 km^2.

The Amtkeli River basin. Amtkeli River originates in the south-western slope of the Chkhalta range. An old vast cirque of the northwestern direction is well represented in its headwaters. Today, there is only one small cirque type glacier of the northwestern exposition. In 1960, its area was 0.08 km^2, and by the 2014 data—it is 0.07 km^2.

3.2.4 Glaciers of the Enguri River Basin

The Enguri River basin embraces the western section of the Central Caucasus, northern slope of the Svaneti range and northern and western sections of the Samegrelo range. There are located the high peaks such as Ushba, Gistola, Tetnuldi, Bashili, Lakutsa, Shkhara, Tikhtengeni, etc., the heights of which exceed 4500 m.

The Enguri River basin is the largest in Georgia according to the number and area of modern glaciers (Fig. 3.16). It exceeds all other basins combined together. Such distribution of modern ice cover entails the theoretical and practical importance of its study.

The study of the glaciers of the Enguri River basin has begun in the nineteenth century; famous researchers of the Greater Caucasus, such as Abich (1865), Dinik (1890), Podozerskiy (1911), Reinhardt (1936), Rutkovskaya (1936), Maruashvili (1937), Tsereteli (1959), Ivankov (1959), Kovalev (1961), Khazaradze (1971), Shengelia (1975), Inashvili (1975), Gobejishvili (1989, 1995), etc., gave us the information about glaciers.

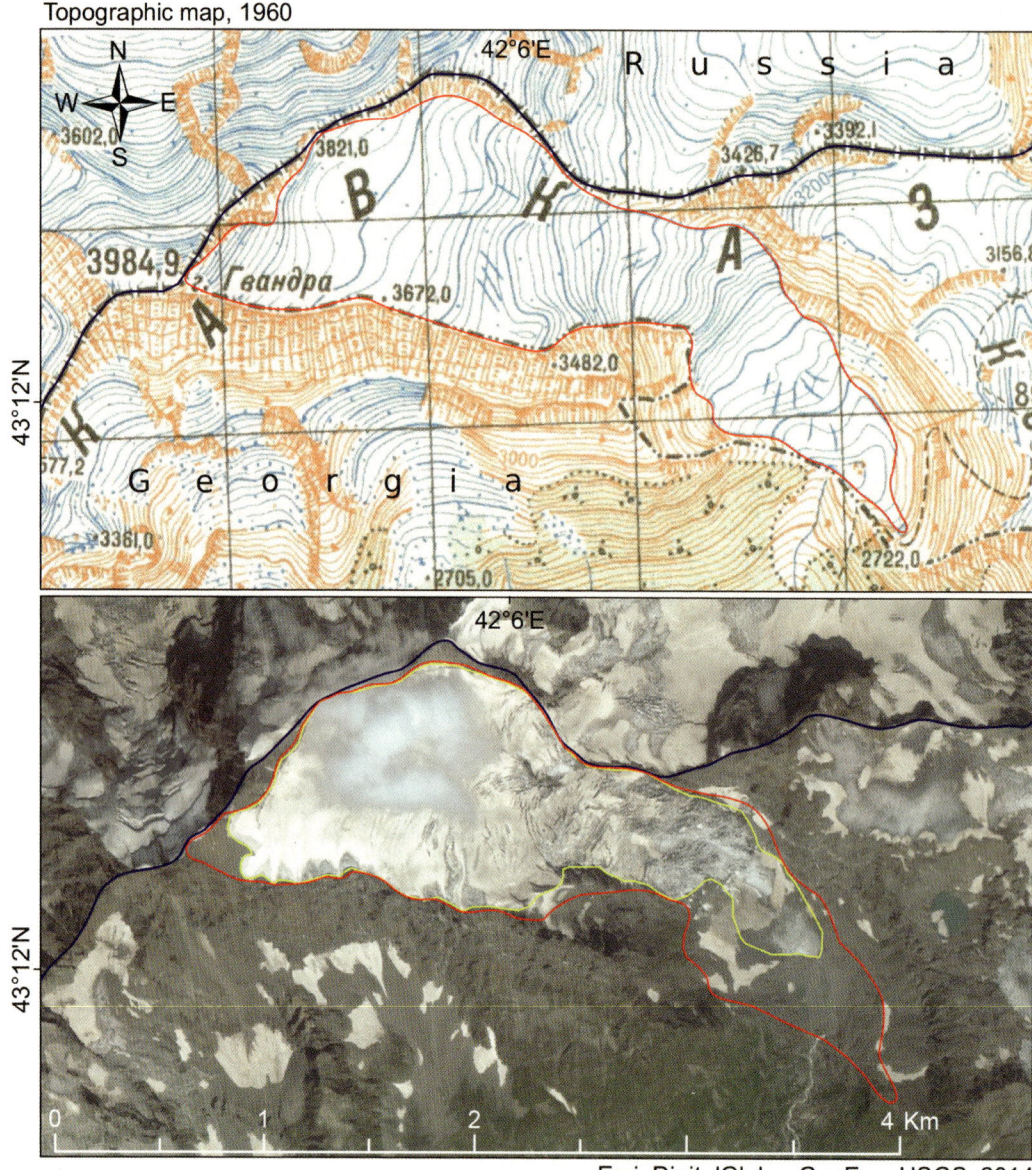

Fig. 3.15 Sakeni glacier reduction in 1960–2014

Since 1974 the phototheodolite survey of the valley type glaciers in the Enguri River basin such as: Kvishi, Ladevali, Dolra, Chalaati, Ushba, and Naumkvani (R. Gobejishvili), was conducted. It should be especially mentioned the period of 2007–2014 years, because in this period we conducted annual expeditions in the entire Enguri River basin. Except the mentioned expeditions, we have studied glaciers located there based on the field and a desk processing of the modern aerial images. The comparison of different topographic maps and literature allowed us to express our opinion on glacial dynamics both of entire Enguri River basin and of its individual tributary river basins (Fig. 3.17).

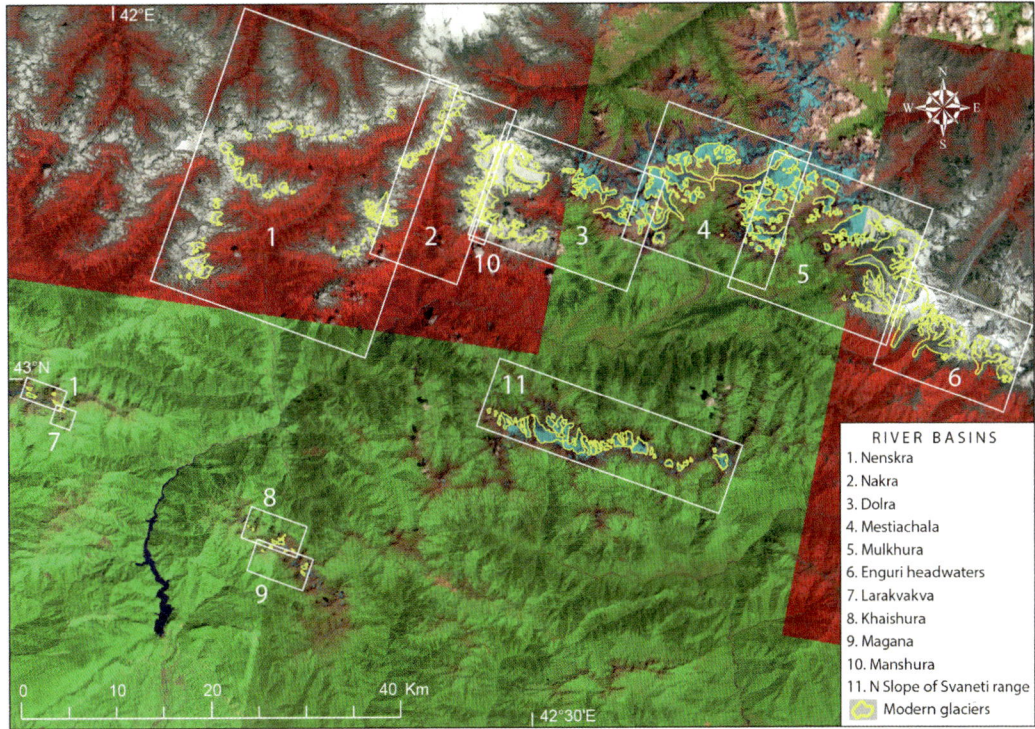

Fig. 3.16 Distribution of modern glaciers of the Enguri River basin according to the tributary river basins

Fig. 3.17 Distribution of the glaciers of the Enguri River basin according to its tributary river basins

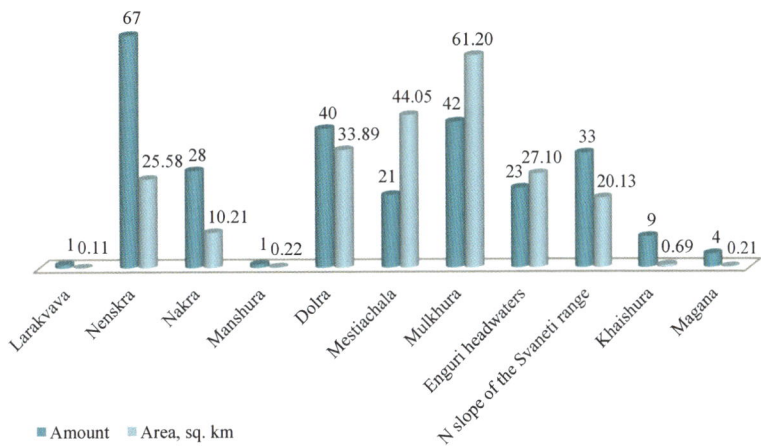

By K. Podozerskiy there were 174 glaciers in the Enguri River basin with the total area of 333.03 ± 4.57 km^2. By the topographic maps of 1960 the number of the glaciers was 299 and the area—323.70 ± 5.72 km^2. Increase in number of the glaciers in this period was stipulated by their division during the glaciers' retreating and by the fact that not all the glaciers were marked on the old topographical maps, but the reduction in area by $2.80 \pm 1.57\%$. does not correspond to reality. Here the data of Podozerskiy are low, which should be entailed by the incorrect showing of glaciers in the topographical maps of those times (it relates to the firn basins) and omitting some of the glaciers. By the data of 2014 here are 269 glaciers with the total area of 223.39 ± 4.6 km^2. In 1960–2014 years the number of glaciers decreased by 30 (10.03%) in the Enguri River basin, while the area reduced by ~ 109.08 km^2 (32.78 ± 1.57%).

In the Enguri River basin according to the number the small glaciers occupy the first place, the areas of which do not exceed 0.5 km^2. According to the area the large glaciers occupy the leading place.

It should be noted that there such three glaciers the area of which exceeds 10.0 km^2. It is Lekhziri—23.26 ± 0.35 km^2, southern Tsaneri—12.60 ± 0.18 km^2, and northern Tsaneri—11.53 ± 0.14 km^2. By the data of 1960 there were eight glaciers in the Enguri River basin.

Then come the glaciers, the area of which is over 8.0 km^2. These are—Kvitlodi—9.76 km^2, Adishi—9.50 km^2, Khalde—8.78 km^2, and Chalaati—8.59 km^2.

There are also 3 glaciers, the area of which exceeds 5.0 km^2, they are: Kvishi—7.45 km^2, the central Lekhziri—6.27 km^2 and Dolra—5.48 km^2.

The total area of all above listed glaciers is 103.22 km^2 and their share in the total area of the glaciers of the Enguri River Basin is 46.20% and in the total are of the glaciers of Georgia—29.01%.

According to the number the cirque glaciers dominate in the Enguri River basin (Fig. 3.18). Their share in the total area of the basin's glaciers is 36.80%. Then come the hanging glaciers (21.93%), valley (19.33%) and cirque-valley glaciers—(12.63%).

According to the area the valley glaciers are the first (52.76%). Then come the compound-valley glaciers (23.24%). There are 6 glaciers in total in the Enguri River basin glacier (western Ushba, southern Ushba, Chalaati, Lekhziri, Khalde, and Shkhara). It should be also noted that there are seven compound-valley glaciers in total on the southern slope of the Greater Caucasus, six of which are located in the Enguri River basin and 1—in the Rioni River basin (Buba glacier).

According to the area of the smallest territory occupy the cirque-hanging glaciers.

Due to the complex and highly fragmented conditions of the Enguri River gorge there are the glaciers of all expositions (Fig. 3.19). The glaciers of northern (49 glaciers) and southern expositions (49 glaciers) are distributed evenly according to their number. Although the main center of the glaciations is the southern slope of the main watershed range, the first place occupy the glaciers of overall northern exposition (N, NE, NW), they occupy 44.60% of the total amount of the glaciers of the Enguri River basin. The glaciers of overall southern exposition occupy the second place.

The large glaciers of the Enguri River basin are located mainly in the southern slope of the Greater Caucasus and it is natural that their exposition is overall southern. That is why, according to the area the first place occupy the glaciers of southern exposition; they occupy 53.77% of the total amount of the glaciers of the Enguri River basin.

As for the glaciers of overall northern exposition, they are mainly located in the branch

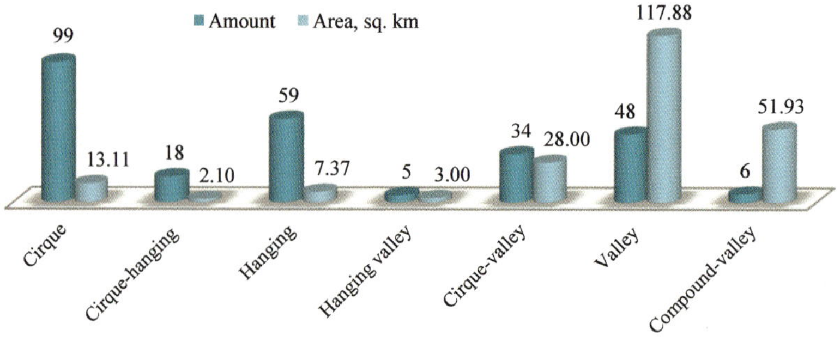

Fig. 3.18 Distribution of the glaciers in the Enguri River basin according to the morphological types

Fig. 3.19 Distribution of the glaciers in the Enguri River basin according to the exposition

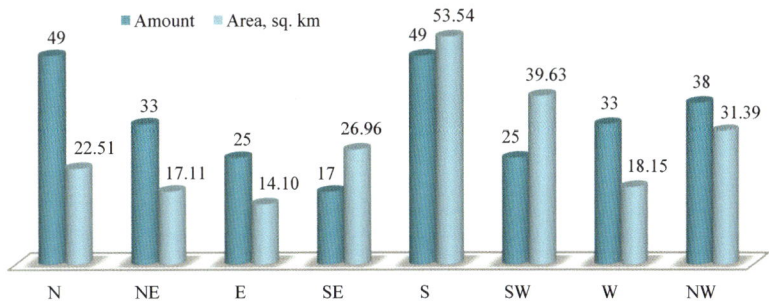

ranges of the Greater Caucasus, the direction of which is latitudinal or submeridional and are characterized by small areas of the glaciers. That is why the glaciers with the overall southern exposition occupy the first place.

Modern glaciers are distributed unevenly in the Enguri River basin not only by the individual orographical units but by the individual tributary river bans as well. To have clearer idea, let us consider the single basins separately.

The Larikvakva River basin is located in the southern slope of the Khojali massif, it is the first right-hand tributary of the Enguri River, where there is a small, cirque glacier of southern exposition. Its area is 0.11 km^2.

The Nenskra River basin is the largest by its area among the Enguri River tributaries. It occupies ∼625 km^2. 24.90% of the total amount of the glaciers of the Enguri River basin is the share of this basin and occupies the first place in the Enguri River basin in this regard. By the area of the glaciers it is behind the Mulkhura, Mestiachala, and Dolra River basins and the Enguri headwaters. Morphometric and morphographical conditions of the relief of the Nenskra gorge cause the uneven distribution of the glaciers here. The Valley type glaciers are mainly found in the right-hand tributary basins of the Nenskra River and on the north\western slope of the Shdavleri range. There are small cirque glaciers on the southern slope of the Greater Caucasus (Tielidze et al. 2015d).

There were 54 glaciers in the Nenskra River basin by K. Podozerskiy with the area of 50.54 km^2. By the data of the topographical maps of 1960 the area of the glaciers was 48.62 km^2 and their number—75. Such

variability of the glaciers was stipulated by the disappearance of small glaciers on the one hand and on the other—by their division during the valley glaciers retreating. And by the data of 2014 there are 67 glaciers with the total area of 25.58 km^2.

The morphology of the relief of the Nenskra River gorge causes the existence of numerous small cirque glaciers here. These types of glaciers occupy 44.77% of the total number of the glaciers of the entire basin. Then come the glaciers of cirque-valley (22.38%) and valley types (19.40%). We have the vice versa situation regarding the area occupied by them. Valley glaciers occupy 55.98% of the total area of the glaciers of the Nenskra River basin, cirque-valley glaciers—25.99% and cirque glaciers—14.69%. The share of the other morphological types of glaciers is insignificant (Fig. 3.20).

The glaciers of the Nenskra River gorge are mainly located in the Kharikhra and Shdavleri ranges, which are of submeridional directions, therefore, here prevail the glaciers of overall northern exposition both by the number and area; they occupy 53.73% of the total number and 59.81% of the total area of the glaciers of the entire basin (Fig. 3.21).

Kharikhra and Shdavleri glaciers are distinguished in the Nenskra River basin y their morphological and morphometric features and dimensions.

The Shdavleri glacier is a valley type glacier of western exposition, its length is of 3.32 km and the area—2.31 km^2. In 1960 its area was 2.48 km^2 (Fig. 3.22). The glacier starts from the two independent firns, which are located on the northern slope of the Mount Shdavleri (3994 m).

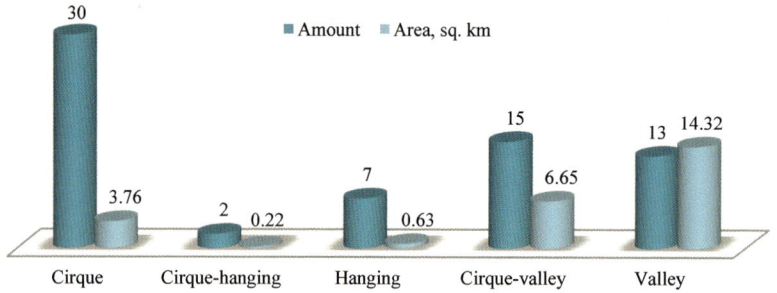

Fig. 3.20 Distribution of the glaciers in the Nenskra River basin according to the morphological types

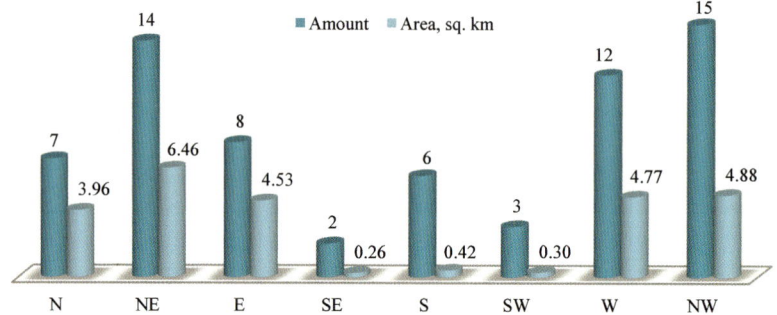

Fig. 3.21 Distribution of the glaciers in the Nenskra River basin according to the exposition

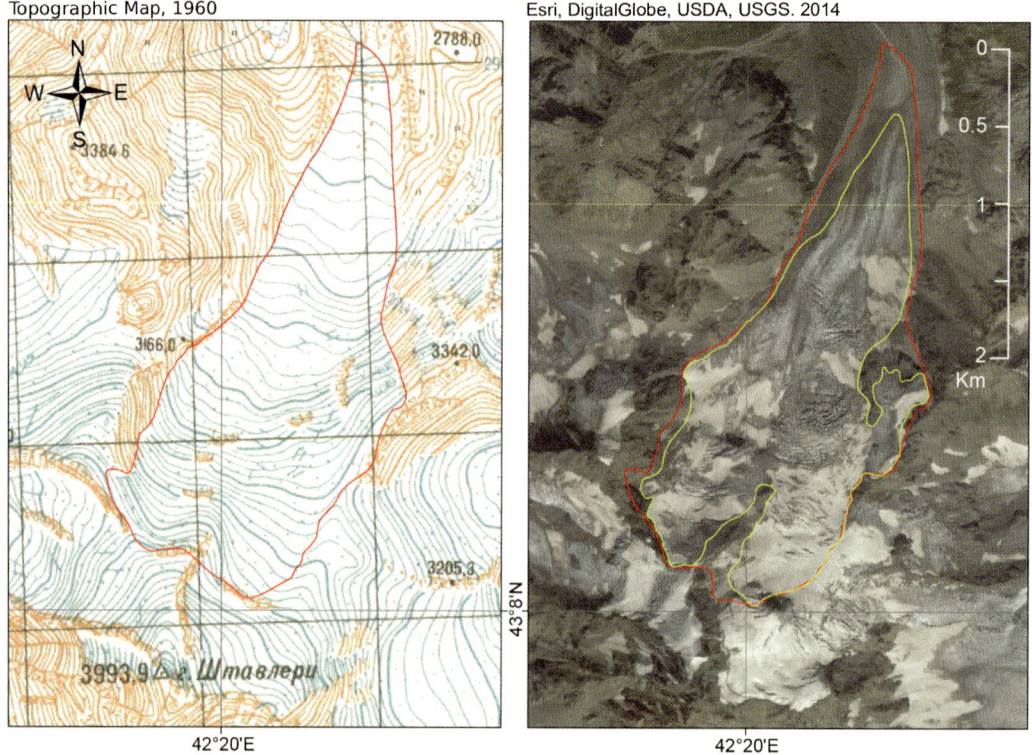

Fig. 3.22 Shdavleri glacier reduction in 1960–2014

The eastern firn is connected to the firn of the glaciers of the Nakra River. Two icefalls are developed at a whole length of the Shdavleri glacier, one—when flowing out from the firn and another—in the middle part of the tongue. The ice tongue is covered with the thin weathered material and is well expressed in the relief. The tongue ends by a pointed form. The lateral stade moraines well expressed on the both sides of the tongue protect it from contamination; inside the stade moraines there can be found the well-expressed micro-stade moraine by which it is possible to identify the parameters of the glacier retreating. By the data of 2014 its ice tongue ends at a height of 2730 m above sea level.

The Nakra River basin is one of the smallest in the Enguri River basin. Its area is ~ 150 km². The Nakra River basin is of meridional direction. To the west it is bounded by the Shdavleri range, to the east—by the Kvishi range (height of some peaks exceed 3900 m here), while to the north it is bounded by a small part of the Greater Caucasus watershed range.

By K. Podozerskiy there were 26 glaciers Nakra River basin with a total area of 20.24 km². By the data of 1960 there were 31 glaciers with the area of 18.49 km². Increasing in the number of glaciers and reduction in their area within the mentioned two periods were well subjected to the increase in number of the glaciers in the first part of the twentieth century in parallel with the reduction in the total area of the glaciers, but the picture is different in the last 54—year period. By the data of 2014 there are 28 glaciers in the basin with a total area of 10.21 km². During the last 54 year the number of glaciers was decreased by 3 and the area was reduced by 44.79%.

In the Enguri River basin the Nakra River basin is behind the rivers of Nenskra, Mulkhura, and Dolra and the northern slope of the Svaneti range by the number of glaciers.

The ratio of the number and area of the glaciers indicates that there are basically the small cirque glaciers in this gorge. The glaciers are distributed by morphological types and exposition as follows (Figs. 3.23 and 3.24).

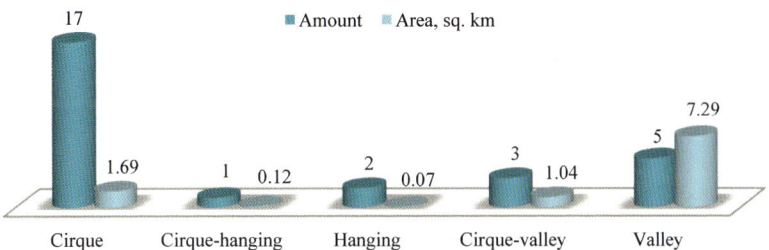

Fig. 3.23 Distribution of the glaciers in the Nakra River basin according to the morphological types

Fig. 3.24 Distribution of the glaciers in the Nakra River basin according to the exposition

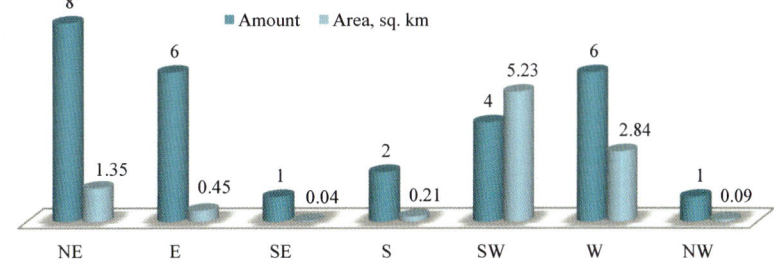

The largest glaciers of the basin—the Nakra and Leadashti are located in the western slope of the Kvishi range.

The Leadashti glacier is the largest glacier in the Nakra River basin with an area of 3.47 km^2. It is a valley type glacier and has an extensive firn field; the glacier tongue is clean and after flowing from the firn ends at the ledge. Its length of 4.02 km. The ice tongue ends at a height of 3170 m above sea level. The firn exposition is southern, while the lower section of the firn and the tongue are of western direction. Due to grandiose ledge the glacier does not have the moraines. In early times, the ice tongue had a form of an icefall and the loose material was collected at the bottom and ledge. The ice tongue overflows from the top of the ledge still today at a short distance. In 1960, the area of the Leadashti glacier was 4.29 km^2. Glacier area was reduced by 19.11% in the years of 1960–2014 (Fig. 3.25).

As for the Nakra glacier, its area was 2.02 km^2 in 1960 and 1.42 km^2 is in 2014. During this period its area was reduced by 29.70% (Fig. 3.25).

The Manshura River basin is located on the southern slope of the Kvishi range. There is a small cirque glacier of southwestern exposition with the area of 0.22 km^2. In 1960, there were two small cirque glaciers in the basin with the total area of 0.48 km^2.

The Dolra River basin is located on the southern slope of the Greater Caucasus watershed mountain range. The Dolra gorge has a trough form at a whole length. There is a well-preserved cirque in its headwaters, which is originated due to impact of glaciations in Late Pleistocene. This vast cirque was a feeding area of the old glaciation in Dolra. United ice tongue flowed in the Enguri gorge in Wurm.

In the Dolra River basin the glaciers are located on the eastern slope of the Kvishi range,

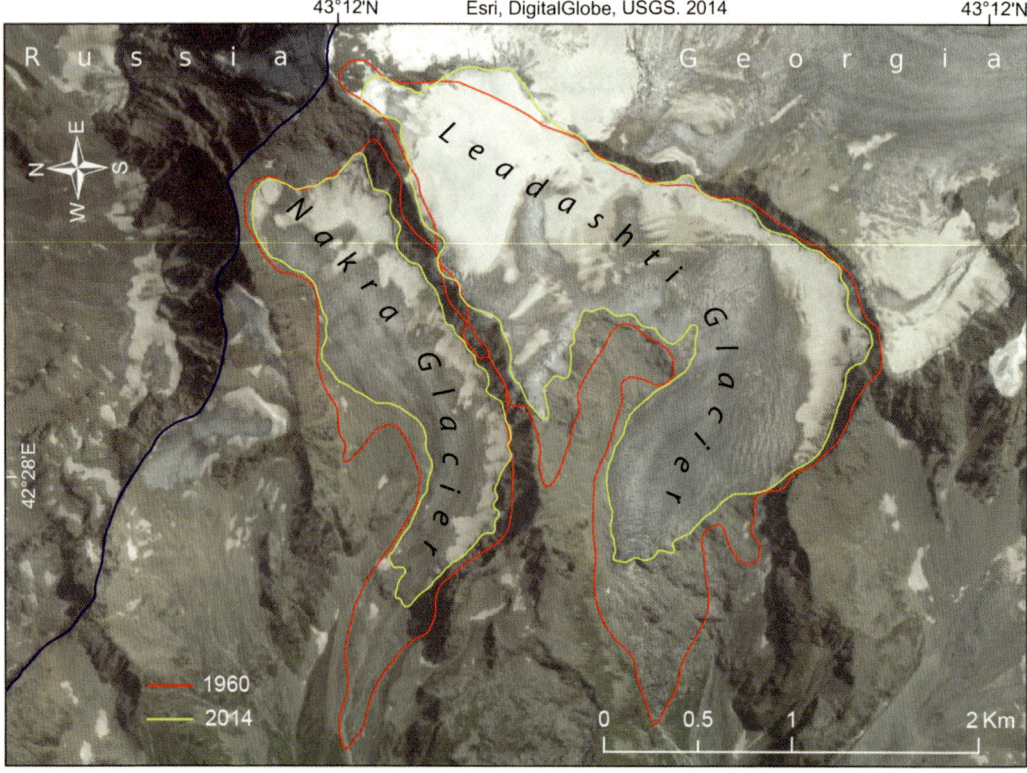

Fig. 3.25 The reduction of Leadashti and Nakra glaciers in 1960–2014

on the northern slope of the Baki range and on the southern slope of the Greater Caucasus between the Mount Donghuzorun (4454 m) and Mount Ushba (4700 m) (Tielidze et al. 2015e). By K. Podozerskiy there were 16 glaciers with the area of 48.51 km^2. By the data of 1960 the number of glaciers has increased up to 28 but their areas almost have not been changed— 48.60 km^2. When we compared the topographical maps of the both periods, the analysis showed that the firn valleys are very incorrectly depicted on the old topomaps. Namely, it concerns to the largest valley glaciers Kvishi, Dolra, and Ushba. There are 40 glaciers there by the data of 2014 with the total area of 33.89 km^2. In the years of 1960–2014 the number of the glaciers in the Dolra River basin increased by 12 and the area decreased by 30.26%.

Share of the glaciers of the Dolra River basin is 14.86% of the total number and 15.17% of the total area of the glaciers of the Enguri River basin.

Valley type glaciers create main background for glaciation in the Dolra River basin. The Kvishi glacier with the area of 7.45 km^2 and the Dolra glacier with the area of 5.48 km^2 are distinguished by their special sizes. 38.15% of the total area of the glaciers of the Dolra River basin is the share of the above-mentioned glaciers. The compound-valley glaciers occupy the second place by their area (7.74 km^2). There are only two glaciers of such type (northern Ushba and southern Ushba). Glaciers of other morphological types occupy small area (Fig. 3.26).

In the Dolra River basin, the glaciers of overall southern exposition prevail by their number and area, which are mainly distributed in the Baki range of latitudinal direction (Fig. 3.27).

Detail descriptions of the glaciers of the Dolra River basin have Rulovskaia (1936), Kovalev (1961), Khazaradze (1971), Gobejishvili (1995) and others. We will briefly characterize some of the main glaciers.

The Kvishi glacier. The glacier by this name is given as a joint glacier in the old topographical map. According to the data of 1911 the Kvishi glacier was the compound-valley largest glacier in the Dolra basin, with an area of 34.3 km^2. It was formed by joining several powerful glaciers and was a glacier of the compound-valley type. During the expedition in 1977 (R. Gobejishvili), the Kvishi glacier was already split and was consisted of four independent glaciers—Kvishi, Ladevali, Tsalgmili, and Lakra. In the aerial image of 2014 it is well seen that there are already five glaciers there. This is caused by the fact that after 1977 due to the melting the Tsalgmili glacier was divided into two parts (northern Tsalgmili and the southern Tsalgmili). The ice tongues of the Kvishi basin glaciers are separated from each other by ∼0.3–0.5 km and experience intensive retreating (Fig. 3.28). It should be noted that if at early times the Kvishi glacier tongue was covered by a very thick loose material (this factor that has always had a large dead glacier), today the ice tongues of the individual glaciers are characterized by nearly pure surfaces.

Fig. 3.26 Distribution of the glaciers in the Dolra River basin according to the morphological types

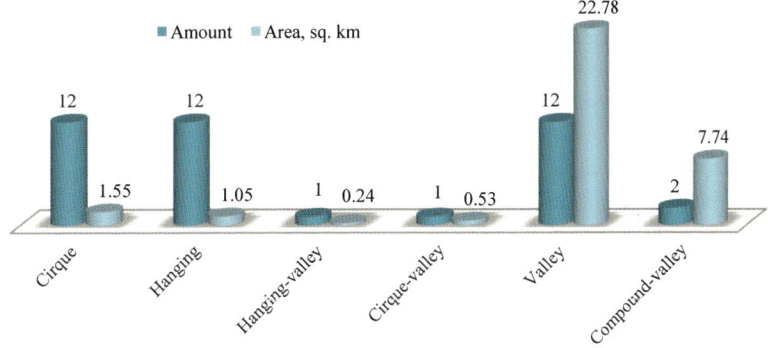

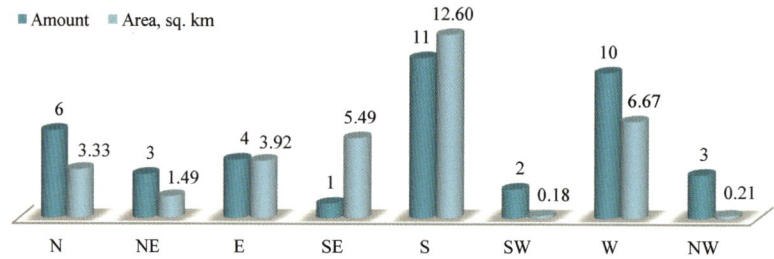

Fig. 3.27 Distribution of the glaciers in the Dolra River basin according to the exposition

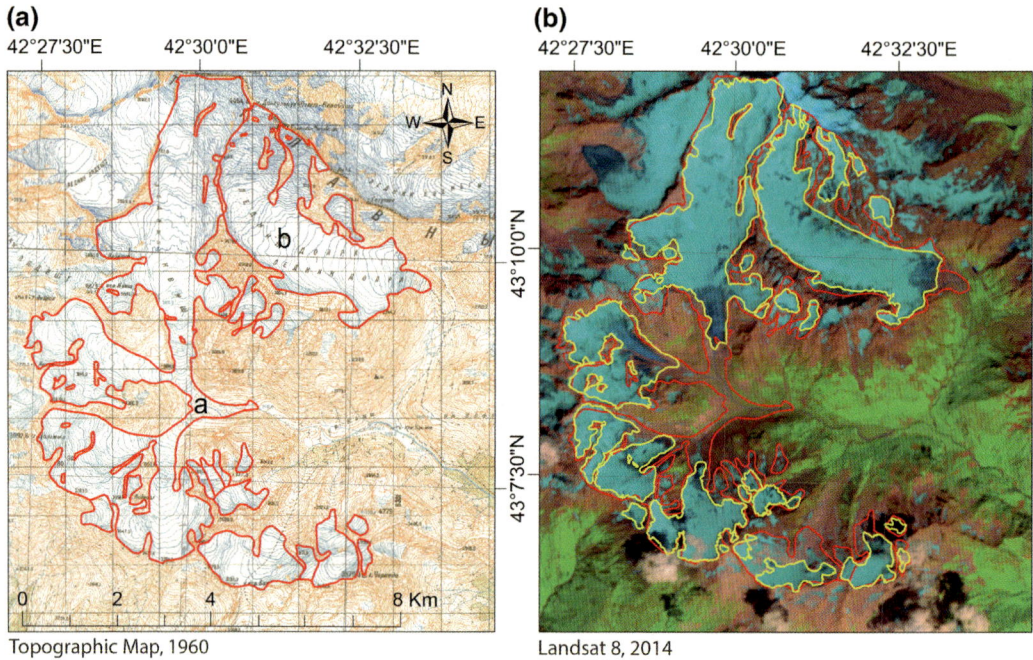

Fig. 3.28 The reduction of Kvishi (**a**) and Dolra (**b**) glaciers in 1960–2014

The Dolra glacier is a valley glacier of southeastern exposition. In 1960 its area was 8.01 km^2 and length—6.32 km. In 2014, its area is 5.48 km^2 and an length—5.68 km. Its shortening in length is caused by the fact that in 1960 the glacier was ended with the ice tongue hanged from the ledge, and it experienced strong mechanical destruction. However, compared to the other glaciers of the Greater Caucasus, the big indicator of retreating in the Dolra glacier has not been recorded (Fig. 3.28). This story is one factor may be that the Dolra ice tongue ends at the highest elevation—2940 m above sea level as

compared to the glaciers of about the same size in the Enguri River basin.

Well-expressed lateral stade and micro-stade moraines go along the both slopes of the Dolra River gorge below the ledge, which allows us to discuss its dynamics after the LIA maximum.

The Ushba glacier contour like the Dolra glacier contour is depicted slightly incorrectly in the catalog of K. Podozerskiy; in particular, the firn valley contour of the left flow goes beyond the real firn valley boundaries and includes the larger area. As for the glacier tongue, compared with the ice tongue depicted in the topographic map of 1960, is slightly shorter, which in our opinion is not true. According to the catalog of K. Podozerskiy the glacier area was 11.3 km^2 and it was a compound-valley type of glacier.

The Northern Ushba and Southern Ushba glaciers were presented as one, compound-valley glacier in 1960, its area was 9.50 km^2. The glacier was consisted of four flows. Two of the main left-hand flows were flowed down from the Ushba slopes, and the right-hand two flows—from the Shkhelda slopes. The glacier was of overall western exposition. Its ice tongue was ended at a height of 2400 m above sea level.

We had the last expeditions to the Ushba glacier in 2012, 2013, and 2014, during which we made a marking and visual inspection of the glacier tongue. In order to specify the data we compared the materials obtained during the field survey with the latest airphotos.

Recent data prove that the compound-valley glacier is divided and is presented as two compound-valley glaciers (each consisting of two flows). As it can be seen from the satellite images, its division would have happened in the years of 2000–2005, because in the aerial image of 2000, there can be still observed a little contact, but in the image of 2014, the northern flow is already distanced from the southern flow. Thus, today we have two glaciers. Out of them, the main flow is the southern Ushba anyway; its area is 4.74 km^2, while the area of the northern Ushba is 2.99 km^2. Glacier area was reduced by 18.63% in the years of 1960–2014 (Fig. 3.29).

Three medial moraines go along the ice tongue surface (from the middle section) of the southern Ushba glacier, which are adjacent to each other in the last part of the glacier tongue and cover its surface by a powerful loose material; in spite of it, the medial moraine do not lose their morphological signs due to depressions among them.

Stade moraines of the Ushba glacier are weakly expressed due to the existence of a high ledge there. They are comparatively well presented in the front section of the ice tongue. By the data 2014 the ice tongue ends at a height of 2600 m above sea level.

Water stream flowing from the Ushba glacier overflows the ledge in about one kilometer and creates one of the most beautiful waterfalls in Georgia, the Shtugra waterfall.

The Mestiachala River basin is located on the southern slope of the Greater Caucasus from the Mount Ushba to the Mount Bashili. It is a right tributary of the Mulkhura River. Its length is ∼11 km and the area of the basin is ∼150 km^2. Mestiachala River gorge maintains a trough form, while the bottom of the gorge is transformed due to the erosion processes below Mestia to the village of Latali. However, the trace of the old glaciation is well preserved on the slope. In the Late Pleistocene, the glaciers of the Mulkhura and Mestiachala River gorges were united and with this united tongue they went down to the village of Latali. Chalaati glacier located in this basin comes down to the lowest height among the Caucasus glaciers, to an altitude of 1960 m and intrudes into the forest zone (Tielidze et al. 2015f).

By Podozerskiy there were 10 glaciers in the Mestiachala River basin with the total area of 55.23 km^2. By the data of 1960, there were 25 glaciers with the area of 57.52 km^2. If the increase in number of the glaciers were in harmony with the common regularities, the same cannot be said for the area, because as in the other basins, the area must also be reduced. The reason is easy to explain. Old maps show the firn valleys of the Chalaati and Lekhziri glaciers very inaccurately, which led to increase in the area against the general background of the retreat of glaciers in the years of 1911–1960. By the data of 2014 there are 21 glaciers in the basin with the total area of 44.05 km^2. During the 54-year period the glaciers number has decreased by 4 and their area has reduced by 23.41%.

It should be noted that at present the largest glacier of the southern slope of the Greater Caucasus is located in this basin, it is a compound-valley glacier of Lekhziri. There is a second compound-valley glacier in this basin—Chalaati. These two glaciers create main conditions for glaciation (together with the separate Central Lekhziri). Their share in the area of the glaciers of the entire basin is 86.56%.

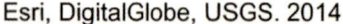

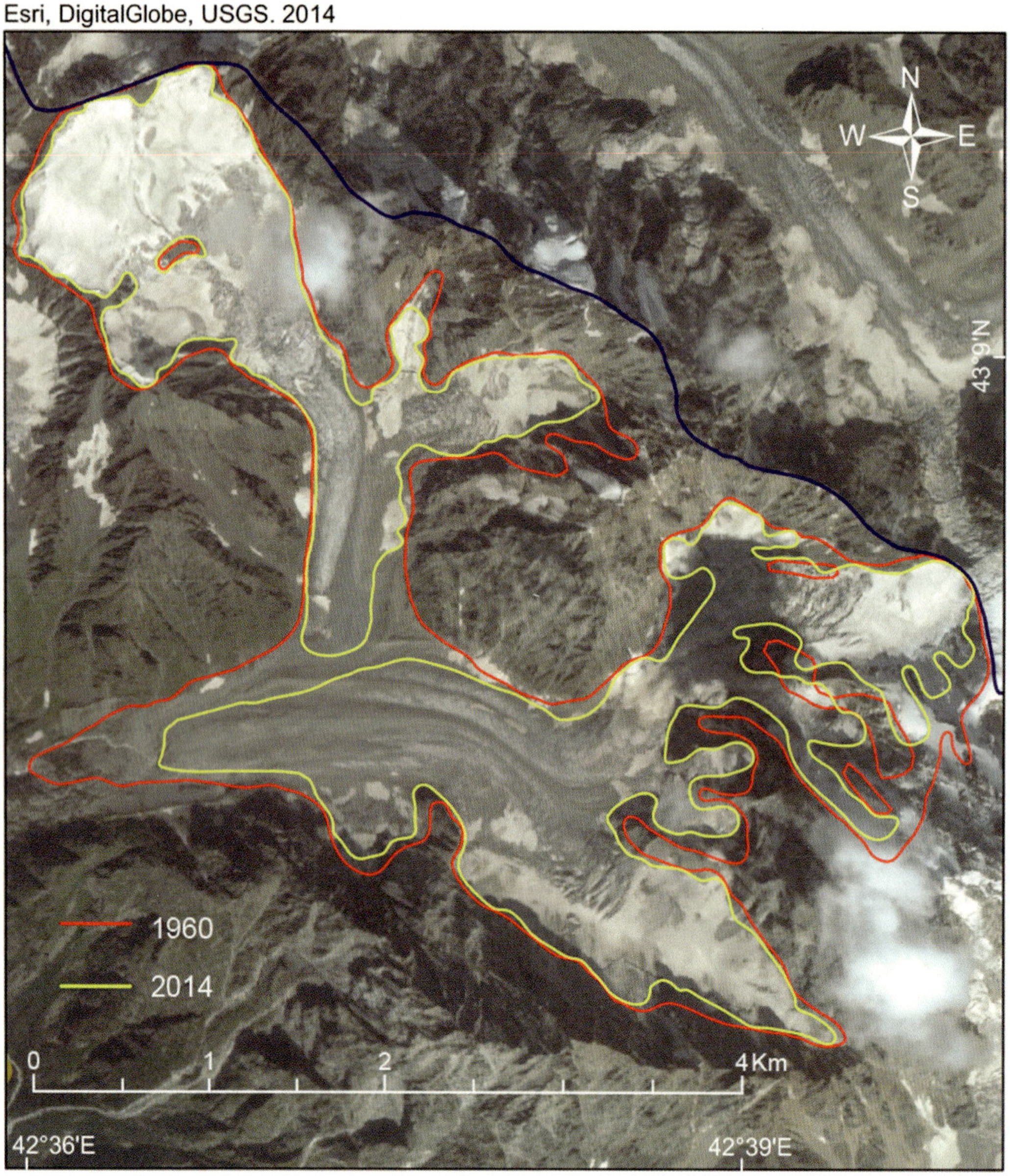
Esri, DigitalGlobe, USGS. 2014
N
W
E
S
43°9'N
1960
2014
0
1
2
4Km
42°36'E
42°39'E

Fig. 3.29 The Ushba glacier reduction in 1960–2014

In the Mestiachala River basin the leading place by the number occupy the small cirque glaciers (Fig. 3.30).

As for the exposition, the glaciers of the overall southern exposition dominate in the Mestiachala River basin according to the number and area. Their share is 71.42% in the number and 90.44% in the area of the glaciers (Fig. 3.31).

The largest glacier of the Georgia—the **Lekhziri glacier** is distinguished by its morpographical and morphometrical features and sizes.

It is a compound-valley glacier and consists of two main flows. According to the data of 1911 the glacier area was 40.84 ± 0.34 km^2 and its tongue flowed down at a height of 1730 m above sea level (Fig. 3.31a). For that time Lekhziri was

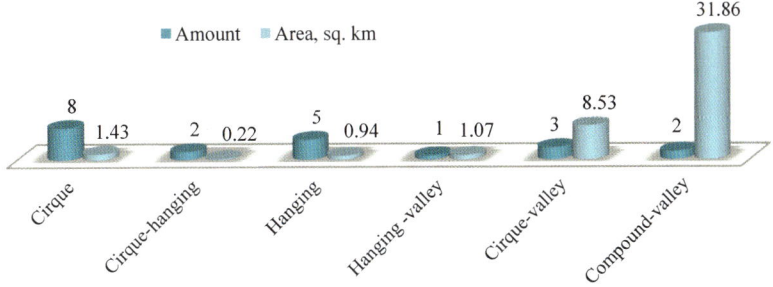

Fig. 3.30 Distribution of the glaciers in the Mestiachala River basin according to the morphological types

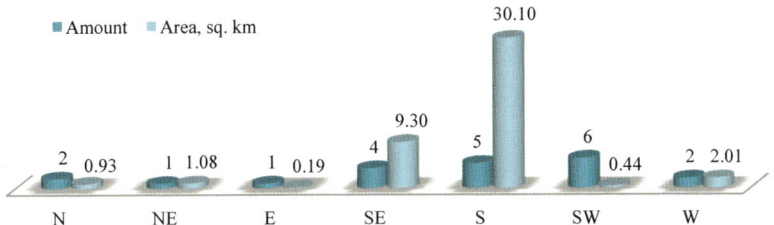

Fig. 3.31 Distribution of the glaciers in the Mestiachala River basin according to the exposition

the third largest glacier in Georgia in size after Tviberi (49.02 km^2) and Tsaneri (together with Nageba—48.90 km^2). In 1960 the glacier was consisted of three flows and it had a cross like shape. The glacier's area was 35.96 ± 0.43 km^2. The ice tongue was ended at a height of 1970 m above sea level (Fig. 3.32b, d).

We had the last expedition to the glacier in 2011. Visual observation showed that the central flow of the glacier had a very weak contact with the two main flows. In the aerial image of 2012 this contact was already split. Therefore, we can conclude that their split took place exactly in 2012. As a result we got the northern Lekhziri (central flow) glacier—the largest cirque-valley glacier in Georgia and Lekhziri glacier (consists of two flows)—the largest compound-valley type glacier in Georgia (Fig. 3.32c).

The mentioned two main flows are covered by the weathered materials from the place of their junction to the end of their joint ice tongue. Evidently, the sharp surface ablation takes place in this section of the glacier as the ice is being diluted and weakened in many places. Contact between the two streams is expected to be interrupted in several years and two independent from each other glaciers will be formed.

Several medial moraines go along the ice tongue surface of the right and left flows, which means that the tongue is formed by joining of several glaciers. It should be noted that in the last section of the ice tongue of the Lekhziri glacier the medial moraines join each other and overlap the ice tongue by a powerful surface moraine, which makes difficult to separate the tongues of the individual flows. But ∼1 km above the ice language the individual flows are easily distinguishable by their morphographical signs and degree of surface pollution.

By the data of 2014 data the area of the northern (central) Lekhziri glacier is 6.27 km^2. The area of the Lekhziri glacier consisted of two main flows is 23.26 ± 0.35 km^2 and the length (of the left flow)—10.73 km. The ice tongue ends at a height of 2320 m above sea level (Fig. 3.32e). In the years of 1960–2014 glacier area was reduced by 17.88%, while the elevation of the glacier tongue increased by 350 m (Fig. 3.32). So far, the exposition of the glacier is the overall southern.

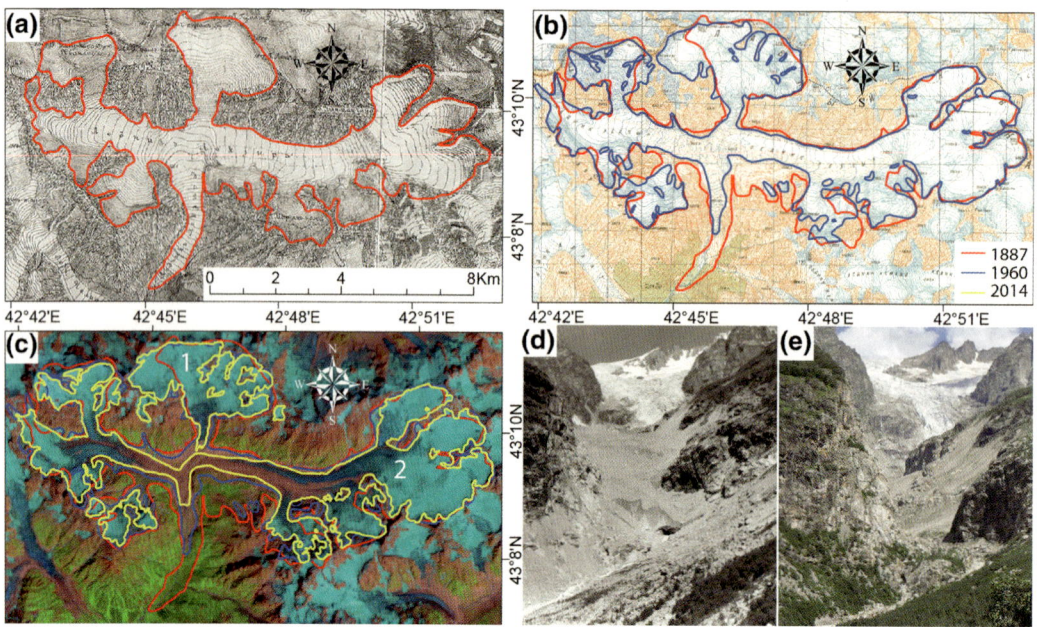

Fig. 3.32 **a** Lekhziri Glacier, topographical map 1887; **b** topographical map 1960, **c** Landsat L8 imagery 2014; **d** Lekhziri Glacier tongue in 1960 (*photo by* R. Gobejishvili); **e** the same view in 2011 (*photo by* L. Tielidze)

The eastern (left) flow of the Lekhziri glacier is formed by the five glaciers, after merging of which the icefall of ∼300 m in height has been developed; further, the ice tongue is characterized by a slight inclination. There are five medial moraine hillocks in its surface, which join each other in the last section of the tongue and are presented by the two powerful medial moraines. In general, the ice tongue is covered by the weathered materials.

There are a lot of different morphosculptural forms in the surface of the ice tongue, in particular, it should be noted the classic manifestation of ogives, which start from the icefall and after ∼1.0–1.5 km they gradually disappear. Their disappearance is caused by the weakening of the ice motion in the glacier in the last part of the ice tongue. There are a large number of glacial tables, wells, the "ant heaps," etc. Lateral moraines are developed to the right side of the before the icefall. Due to the affect of snow avalanches the moraines are not preserved below. However, on the slopes are well-preserved traces of the LIA maximum.

The right-hand western flow of the Lekhziri glacier is consisted of two ice tongues, after their junction there are well-expressed medial moraines on the surface of the glacier. Both of the ice tongues of the flow are covered by loose material in the last section.

The central flow of the Lekhziri glacier is of southern exposition. Its length is 4.08 km and the area—6.27 km². There are the well-preserved lateral and terminal stade moraines in the gorge at a distance of ∼1.5 km from the ice tongue.

The Chalaati glacier. Here is just a small review of the glacier, as below we have considered in details its all the parameters and dynamics of different years (see 4.4. Dynamics of the glaciers in 1960–2014). We would like to say that every year we organize the expeditions to the glacier in 2010–2014. In the summer season of 2011 we conducted a stationary observation for the glacier mass balance research purposes.

Chalaati is a compound-valley glacier and is consisted of two flows and is fed from the slopes of over 4000 m high peaks: Ushba, Chatini,

Kavkasi, and Bzhedukhi. Among the glaciers on the southern slopes of the Greater Caucasus this glacier comes down to the lowest elevation, to 1960 m above sea level, and intrudes into the forest zone. The area of the glacier is 8.59 km^2; the length of the left main flow is 6.90 km.

As it was mentioned, the Chalaati glacier is consisted of two flows. The main is the left flow, three icefalls are developed in its surface, which indicates the existence of the ledge in the under glacier relief. The height of the most powerful icefall is ~300 m and width ~700 m. The lower two icefalls are relatively small. In the vicinity of the icefalls the glacier tongue is rugged by the various fractures (seracs) of the different directions. The sides of the glacier are covered with weathered materials of different thickness. In the contact area of the right and left flows and below the medial moraine is developed and the glacier is covered with a thick layer of weathered material as well. These flows will be split in the near future.

It is possible to identify the drastic change of the Chalaati glacier by comparing of the old and modern images each other (Fig. 3.33).

The Mulkhura River basin is the main center of the Modern glaciations in the Enguri River gorge. Share of the glaciers of this basin is 27.39% in the total area of the modern glaciation of the Enguri River gorge. Glaciers are located in the southern slope of the Greater Caucasus, which is distinguished by the deep fragmentation of the relief and high peaks, with the heights of over 4500 m (Tielidze et al. 2015a).

The last expedition to the Mulkhura basin, particularly, in the Tviberi gorge in 2011. During the expedition, the Italian and Canadian glaciologist-film documentalists conducted the photo and video shooting along the traces of Vittorio Sella. Our joint expedition obtained

Fig. 3.33 Chalaati glacier reduction, in the years of 1890 (*photo by* V. Sella)–2011 (*photo by* L. Tielidze)

(by the field trips and helicopter) very interesting materials of the photo, video and field observations of present morphology and dynamic of the glaciers in this basin.

By the data of K. Podozerskiy there were 11 glaciers in the Mulkhura River basin with the total area of 92.70 km^2. By the data of 1960 there were 31 glaciers and by the data of 2014 there are 42 glaciers with the total area of 61.20 km^2. During the past century, such increase in glaciers number and a sharp reduction in their area is caused by the fact that still in the nineteenth century there was one of the largest glaciers of Georgia—the Tviberi glacier (for more information about Tviberi glacier, see 4.5 Valley glaciers reduction after the Little Ice Age maximum), which was represented as an individual glacier, and the Tsaneri glacier, which was one of the largest glaciers of Georgia. Today, these glaciers are fragmented and are represented as an indvidual simple-valley glaciers.

Share of only six glaciers in the total area of the glaciers of the Mulkhura River basin is 76.84%, their areas are as follows: southern Tsaneri—12.60 km^2, northern Tsaneri—11.53 km^2, Kvitlodi—9.76 km^2, Nageba—4.54 km^2, Asmashi—4.45 km,^{2} and Seri—4.15 km^2, which shows that the basic background for glaciations are made by these glaciers.

The rest of the 36 glaciers are relatively small in size; they are located on the slopes of the major glaciers and end within the altitude of ~2650–3500 m. The ice tongues of the valley glaciers come down to a lower altitude, to ~2140–2480 m above sea level.

According to the morphological types the valley glaciers form the basic background for glaciation in the Mulkhura River basin; they occupy 82.95% of the total area of the glaciers of the basin (Fig. 3.34).

As for the exposition, the first place by the number and area occupy the glaciers of the overall southern exposition; their share is 45.23%

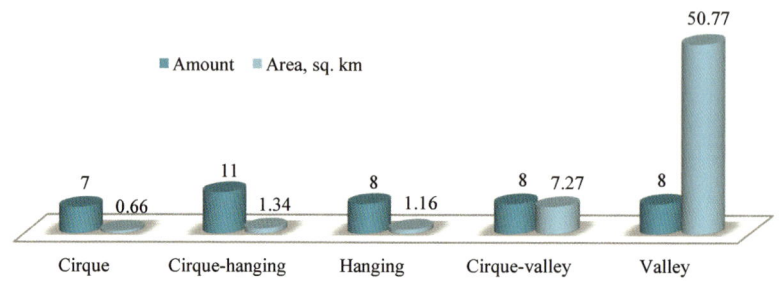

Fig. 3.34 Distribution of the glaciers in the Mulkhura River basin according to the morphological types

Fig. 3.35 Distribution of the glaciers in the Mulkhura River basin according to the exposition

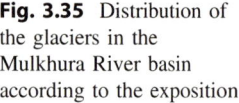

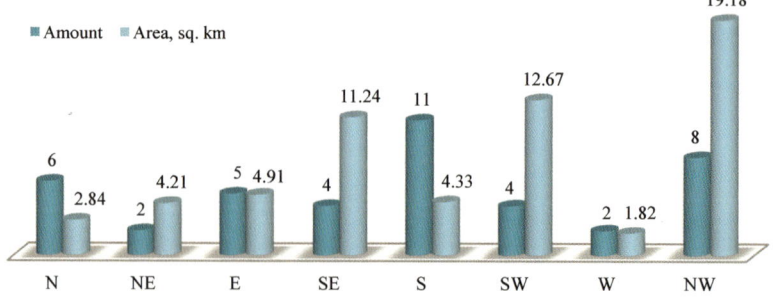

in the total number and 46.14% in the total area of the glaciers (Fig. 3.35).

The Tsaneri glacier with the Nageba glacier was the second glacier in Georgia by its size after the Tviberi glacier in the 80s of the nineteenth century and its area was 48.90 ± 0.49 km^2. Division of the glaciers took place at the end of the nineteenth century. In 1960, the Tsaneri glacier still was the compound-valley type of glacier and its area was 28.3 km^2. For this period, the Tsaneri glacier was the second glacier in Georgia by its size after the Lekhziri glacier. The mentioned glacier now is presented in the form of two glaciers (northern Tsaneri and southern Tsaneri), which descend from the firn valleys that are independent from each other.

The aerial images of 1986 show that these two flows have little contact with each other. And the aerial image of 2000 shows that their contacts are split and the northern flow is quite far away from the southern flow. Their division is likely to be happened in the years of 1986–1990.

Their firns are separated from each other by a branch of the Greater Caucasus mountain range. The ice tongues were joined at 2750 m above sea level and a uniform ice tongue of western exposition was created, which was ended at an altitude of 2380 m. Glacier area was of 28.28 km^2.

The Southern Tsaneri is the second largest glacier in Georgia with the area of 12.60 ± 0.18 km^2 and the length of 8.62 km (Fig. 3.36b). Glacier surface cracks are weakly developed. The ice tongue surface is rich in ablation forms (wells, glacier tables, sun cups, grooves made by melted water, etc.). The glacier tongue ends at an altitude of 2525 m above sea level. The well-expressed lateral stade moraines stretch along to its both sides.

Georgia's third largest glacier is the **Northern Tsaneri** with the area of 11.53 ± 0.14 km^2. It is a valley glacier and is of southwestern exposition. The last part of the ice tongue is hanged over the ledge and it ends at an altitude of altitude of 3020 m above sea level (Fig. 3.36a).

The Nageba glacier is a valley glacier of northwestern exposition. The glacier firn starts from the Tetnuldi Peak at a height of 4858 m and is surrounded by the Lakchkhilda and Kvarasha ranges. The glacier tongue creates the icefall with the cracks of different direction after flowing out from the firn. By the data of 2014, the glacier area is 4.54 km^2. Its tongue ends at the height of 2800 m (Fig. 3.37).

Well-expressed lateral stade moraines stretch along to the both sides of the glacier, which join the stade moraine of LIA Maximum of the Tsaneri glacier. This fact indicates that the glaciers were combined and represented one glacier during the last stade glaciation. There are several short stade moraines into the stade moraines on the slopes and at the base of the valley.

By the data of 1960 the area of the Nageba glacier was 6.07 km^2. Its tongue ended at a height of 2413 m above sea level. During the last 54 years the glacier its tongue was raised by 400 m. Such a drastic change is caused by the fact that in 1960 the last ~ 1.5 km section of the ice tongue was narrow and sharply pointed; accordingly, that section was melted soon (Fig. 3.37).

The headwaters of the Enguri River embraces the Central Caucasus from the Mount Gistola to the Mount Namkvani, or the headwaters of the Enguri River itself and the river basins of its right-hand tributaries—Adishchala and Khaldechala. In the literature, this section of the Greater Caucasus is known as a Bezingi wall. The southern slope profile the Bezingi wall is short and steep, which fully reflects the name of the "Wall" (wooden). There are peaks with the height of over 5000 m.

There are large valley glaciers in the territory within the mentioned frames, such as: Adishi, Khalde, Shkhara, and Namkvani. The river gorges have well-expressed trough forms along their entire length. According to the last years' field studies and the data of aerial images of 2014 we have 23 glaciers there with the total area of 27.10 km^2. This is 12.13% of the total area of the glaciers of the Enguri River Basin.

88.96% of the entire basin area is a share of four main glaciers—Adishi—9.50 km^2, Khalde—8.78 km^2, Shkhara—3.55 km^2, and Namkvani 2.28 km^2. The sizes of the rest of the glaciers do not exceed 1.0 km. By the data of 1960 there were 21 glaciers in the basin with the total area of 34.42 km^2.

Fig. 3.36 Northern (**a**) and Southern (**b**) Tsaneri glaciers in 2012

Morphologically the largest area is occupied by the compound-valley glaciers, while by the number dominate the hanging glaciers. According to the area the valley glaciers are in the second place (Fig. 3.38).

As for the exposition of the glaciers the first place by the number and area belong to the glaciers of the overall southern exposition. There is only one glacier with the western exposition (Fig. 3.39).

The Adishi glacier is one of the beautiful glaciers in the Greater Caucasus (Fig. 3.40), it is divided into three parts: with the firn valley above 3800 m, which is surrounded by the high peaks (Tetnuldi, Adishi, Gistola, Lakutsa, etc.), also with the big icefall (~1000–1300 m in height) and the classic ice tongue. The Adishi glacier is a valley glacier with the southwestern exposition. By the data of 2014, its area is 9.50 km².

The second section of the glacier—the icefall is of grandiose sizes; the glacier tongue flows down to the height of ~1500 m by the angle of ~50° and make visitors admired. The icefall is ~700–800 m in width. It is characterized by disordered cracks. There are many seracs and cliffs there. The icefall is halted at a side of Tetnuldi at a height of ~40–50 m. In this

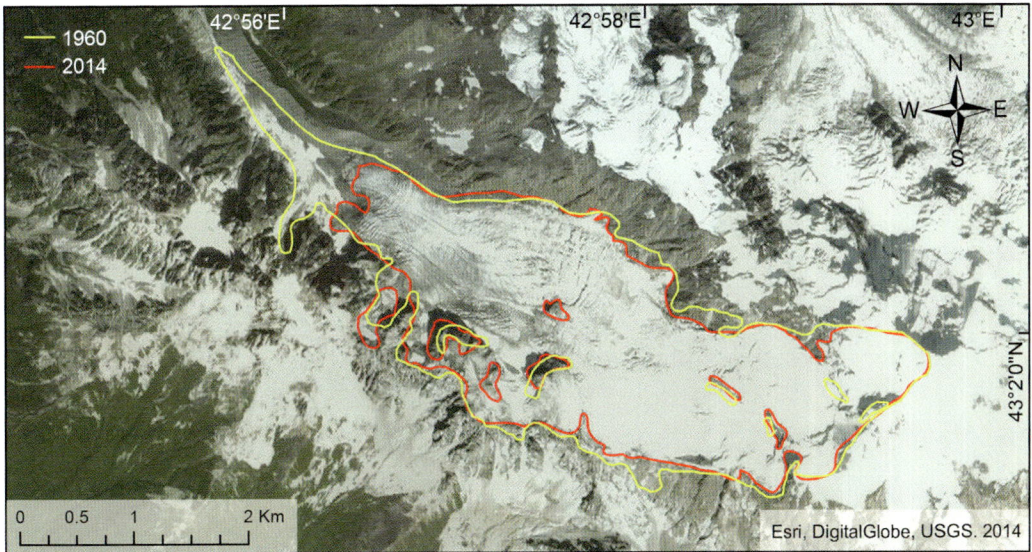

Fig. 3.37 Nageba glacier reduction, in the years of 1960–2014

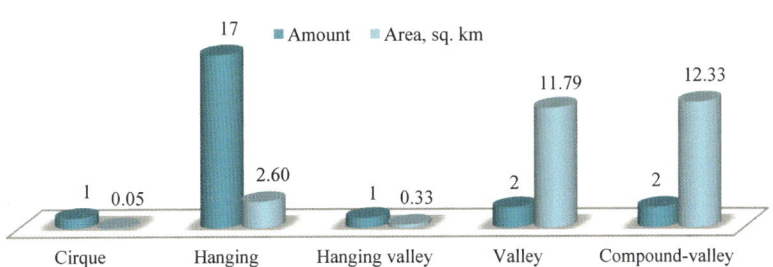

Fig. 3.38 Distribution of the glaciers in the Enguri River headwaters according to the morphological types

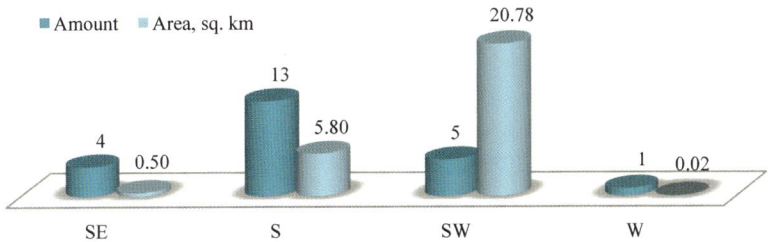

Fig. 3.39 Distribution of the glaciers in the Enguri River headwaters according to the exposition

Fig. 3.40 Adishi glacier reduction, in the years of 1960 (*photo by* Ramin Gobejishvili)–2014 (*photo by* L. Tielidze)

section, the glacier experiences an intense mechanical destruction. Demolished ice accumulates at the base of the icefall and creates a cone of it, which joins the glacier tongue later.

The shape of the glacier changes dramatically from the ice base (2650 m); the tongue slightly inclined (~ 10–$15°$) and in a shape of a beautiful fan ends at the height of 2485 m above sea level. The ice tongue length is ~ 2.5 km. According to the description of 1960 the glacier tongue was covered with a small amount of weathered material, but the last visual inspection and observation of aerial image shows that the amount of weathered material has sharply increased, especially in the sides and in the last part. The tongue is raised in the central part. The ogives stretch along the entire length of it and give a peculiar air to its beauty. The surface of glaciers is wavy, with numerous glacial wells, mills and ablation forms. There are many groove and gorges in its surface created by melted water. Glacier surface devoid of cracks, they are only found in its sides and at the bottom of the icefall.

The well-expressed stade moraines stretch along the both sides of the glacier ice, which make curve after ~ 1.5 km from the end of the tongue and transform into the terminal stade moraine. Microstade moraines are developed on the slope of the stade moraine in front of the glacier and at the bottom of the gorge. Directly in front the ice tongue there can be found several annual curve hillocks (Tielidze et al. 2015b).

The Khalde glacier is a compound-valley glacier with the southwestern exposition. The firn valley is surrounded by the highest peaks of the Greater Caucasus mountain range (Shkhara, Pushkin peak, Jangha, Rustaveli peak, etc.), the heights of which exceed 5000 m. In the upper part it is represented by the three icefalls, which are characterized by a large fracturing. The icefalls are combined at a height of 2800 m. And below, the ice tongue changes its direction to the south-west and ends at a height of 2545 m (Fig. 3.41).

At the place of junction the glacier expands and creates the vast stadium, below of which the glacier flows as one powerful flow. The end part of the glacier tongue is covered with a thick moraine cover, which protects the glacier surface from the intensive melting. That is why in the years of 1960–2014 the average annual retreat of the Khalde glacier is not fixed even as of 10 m/year. This is the lowest figure in this size of glaciers of the Greater Caucasus. There are two medial moraines in the middle of the glacier tongue. Stade moraines stretch along the both sides of the Khalde glacier, which make curves and transform into the terminal moraine. The glacier area is 8.78 km^2.

There can be found a large number of the erratic boulders in the Khaldechala River gorge, of which one—the "Perkhuli" boulder ("boulder of dance"), due to its grandiose sizes is included in the Red Book of Georgia. Such an abundance

Fig. 3.41 Khalde glacier. Google Earth Image, 2013

of large erratic boulders indicates the scale of old glaciation in this gorge.

The Shkhara Glacier is a compound-valley glacier with the southern exposition (Fig. 3.42). It consists of three flows that are combined at a height of 2900 m above sea level. From there the ice tongue is of southwestern direction. The glacier tongue is quite cracked, especially its right section. From the middle part of the glacier's surface is covered by loose material. Its surface is rich in ablation forms ("tables," "ant clusters," etc.). Glacier length (from the central flow) is 3.78 km, and an area—3.55 km². The glacier tongue ends at height of 2540 m above sea level. As the glacier surface is covered with a thick weathered material, the ice melt very weak. Unlike the ice tongue it seems that the melting is quite intensive in the feeding area, because all of the three flows are weakened compared with the year of 1960.

There are well-expressed lateral stade moraines to the both sides of the glacier, which are ended by the terminal moraine.

The glaciers of the northern slope of the Svaneti range. The Svaneti range extends from the Zagaro (Atkveri) Pass to the village of Khaishi; its length is of ∼100 km.

It is a watershed of the Enguri and Tskhenistskali Rivers. It is the first among the Georgia's mountain ranges by modern glacial cover and hypsometrical characteristics. By these features it is divided into three sections: eastern, central, and western. From them, eastern and western sections are hypsometrically lower and the modern ice cover cannot be found there (except of the Mount Dadiashi).

Fig. 3.42 Shkhara glacier. Google Earth Image, 2013

There are the glaciers in the central section, which are located between the sections of Lasili and Leshnuri. Here is the highest peak of the mountain range: the Laila (Laila-Lehli)—4009 m. All the left tributaries of the Enguri River up to Khaishi originate in the Svaneti range. There were 48 glaciers there with the total area of 27.76 km^2 by the data of 1960. By the data of 2014 there are 33 glaciers with the total area of 20.13 km^2.

The largest glaciers of the northern slope of the Svaneti range are formed on the slope of the Laila peak, the Laila peak by itself is covered by a glacier cap (Fig. 3.43). Among these glaciers the largest is the **Eastern Laila**, which is located in the Khumpreri River basin. In 1960, the glacier area was 5.96 km^2; it had the north-western exposition. It was consisted of two flows and it was a compound-valley glacier. Its ice tongue was ended at a height of 2300 m above sea level.

By the data of 2014 the glacier is divided into two parts. From them, the eastern Laila's area is 3.55 km^2 and is twice more than the area of the western Laila (1.24 km^2). As the aerial images show, their division would have happened in about 2000.

The eastern Laila is the biggest glacier in the Svaneti range and it has a northwestern exposition. Its tongue ends at a height of 2640 m above sea level. The glacier length is 4.52 km. Such high indicator of the glacier retreating and rising

Fig. 3.43 The glaciers of the northern slope of the Svaneti range, 2013 (*photo by* L. Tielidze)

its tongue above 340 m is related to the fact that in 1960 the ice tongues was very narrowed at the junction of the glaciers and accordingly, the melting process was going in a high intensity.

The second largest glacier in the northern slope of the Svaneti range is the Lailchala glacier with the area of 2.31 km². It is a valley glacier of northern exposition. Its tongue ends at a height of 2640 m above sea level.

On the northern slope of the Svaneti range the first place by the area occupy the valley glaciers. And according to the number—the cirque glaciers are in the first place (Fig. 3.44).

As the glaciers are located on the northern slope of the Svaneti range, their exposition is the overall northern, only one small cirque glacier is of eastern exposition, which was created as a result of the division of the Lailchala glacier (Fig. 3.45).

The Khaishura River basin is located on the left side of the Enguri gorge between the Svaneti and Samegrelo ranges. Glaciers can be found only on the northeastern slope of the Samegrelo range. There were 12 glaciers in this basin by the data of 1960 with the total area of 1.61 km². By the data of 2014 there are nine glaciers with the total area of 0.69 km². All of the glaciers belong to the cirque type morphologically. As a whole, they have the overall northern exposition. The

mentioned glaciers are located at a height of 2600–3000 m above sea level.

The Magana River is the last left tributary of the Enguri River, where there are modern glaciers in its basin. Magana River basin is located on the southwestern slope of the Samegrelo range. Samegrelo range peaks do not exceed 3150 m, but the mountain range has such orientation toward the Black Sea, that it was possible the development of the small cirque glaciers there (due to the abundance of solid precipitation and morphological conditions of the relief). By data of the 1960, there were 20 glaciers there with the total area of 0.97 km². By the data of 2014 there are four glaciers with the total area of 0.21 km². The glaciers are of cirque type morphologically and they are mainly of northern exposition. Out of four glaciers three have the overall northern exposition and one—southern exposition.

3.2.5 Glaciers of the Khobistskali River Basin

The Khobistskali River originates from the southern slope of the Samegrelo range. Although the heights of the peaks of the Samegrelo range do not exceed the 3150 m, still there are the

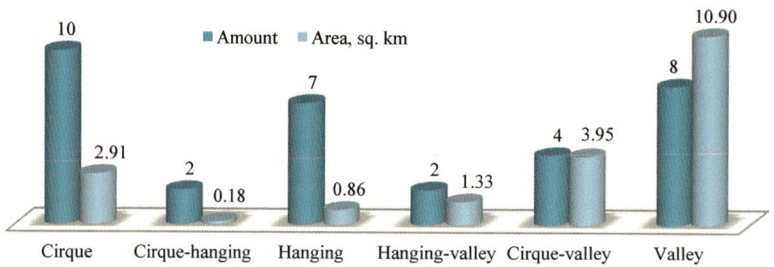

Fig. 3.44 Distribution of the glaciers of the Northern slope of the Svaneti range according to the morphological types

Fig. 3.45 Distribution of the glaciers in the Northern slope of the Svaneti range according to the exposition

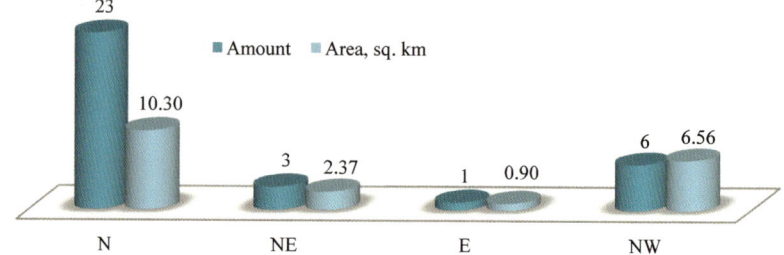

glaciers of small size there. By the data of 1960 there were 16 glaciers in this basin with a total area of 1.12 ± 0.07 km^2. By the data of 2014 there are nine glaciers there with a total area of 0.46 ± 0.03 km^2. Formation of the modern glaciers is determined by: (1) the morphological features of the relief and slope exposition; (2) favorable location towards the Black Sea, which determines the abundance of solid precipitation during the winter period; and, (3) Blowing away of solid precipitation and their accumulation in the deep-seated negative forms of the relief.

Glaciers in this basin are of cirque type. According to the exposition, the two glaciers are of eastern exposition, and the rest of the seven glaciers are of overall northern exposition.

3.2.6   Glaciers of the Rioni River Basin

The Central Caucasus watershed range is a main center of the glaciation in the Rioni River basin from the Mount Namkvani to the Mount Kozikhokhi. There are located such massifs, the height of which exceeds 4000 m. There are the

separate glaciation centers in the segments of Svaneti, Lechkhumi, Shoda-Kedela and Leti ranges, the altitudes of which exceeds 3500 m.

The Rioni River basin is behind only the basins of the Enguri and Kodori Rivers by the number of modern glaciers in the southern slope of the Greater Caucasus; by area it is behind only the Enguri River basin. Modern glaciers are widespread within the study area (Fig. 3.46). Study of their dynamics is of great importance not only in terms of regional research, but also in terms of the general modern glaciation.

Glaciers are relatively well studied in the Rioni River basin. We find many references about them in many papers of Dinik (1890), Rashevskiy (1904), Podozerskiy (1911), Vardaniants (1930), Reinhardt (1936), Tsereteli (1959, 1966), Ivankov (1959), Shengelia (1975), Inashvili (1975), Gobejishvili (1995), etc.

Since 1968 on the southern slope of the Racha Caucasus the glacial-geomorphological studies were conducted with the participation and guidance of R. Gobejishvili. By using the phototheodolite method the survey of the largest glaciers, such as Zopkhito, Laboda, Brili, Chasakhtomi, Kirtisho, Buba, Boko, Tbilisa, and Koruldashi, has been conducted.

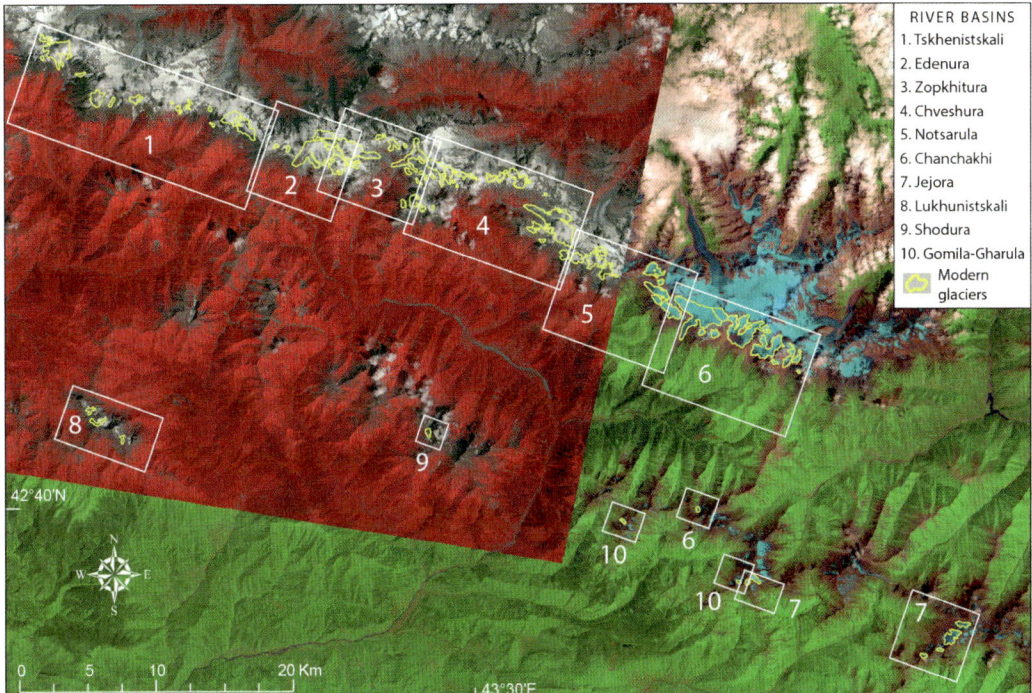

Fig. 3.46 Distribution of modern glaciers of the Rioni River basin according to the tributary river basins

We have been intensively conducting the annual expeditions to all of the river basins of the Racha Caucasus within the years of 2005–2014. By using the field survey and the latest aerophoto images, we studied and identified a number of the modern glaciers, their dynamics and morphological characteristics. We specified the boundaries of the old glaciations as well.

By the data of K. Podozerskiy there were 85 glaciers in the Rioni River basin with the total area of 78.12 ± 1.61 km^2. According to the topographic maps of 1960, there were 112 glaciers in the Rioni River basin with the total area of 76.77 ± 1.66 km^2.

As we can see, during the period of 1911–1960 the area of the glaciers reduced by $2.50 \pm 2.11\%$, and the number—increased by 27. Such unimportant reduction in area was caused by the fact that many of the small glaciers are not shown in the old maps, and the firns of some of the glaciers are depicted incorrectly. The same cannot be said for the period of 1960–2014, because during this period along with the area the

number of the glaciers was also decreased. For today, there are 97 glaciers in the Rioni River basin with the total area of 46.65 ± 1.15 km^2. During the last 54 years the 15 glaciers have been melted at all, while the area has been reduced by $37.86 \pm 2.32\%$.

The Kirtisho glacier is the largest glacier in the Rioni River basin with the area of 4.41 ± 0.07 km^2. It is the valley glacier of the northwestern exposition.

The glaciers are unevenly distributed in the Rioni River Basin not only by the orographic and hypsometric units, but according to the individual tributary river basins as well (Fig. 3.47).

Leading position by the number occupy the cirque glaciers from the morphological types of glaciers distributed in the Rioni River basin. Their share is 34.02% in the total number of the glaciers of the basin. There are a lot of hanging glaciers there—26.80%. Cirque-hanging glaciers are on the third place –14.43%. The cirque-valley and valley glaciers are distributed by equal amounts (Fig. 3.48).

Fig. 3.47 Distribution of the glaciers of the Rioni River basin according to its tributary river basins

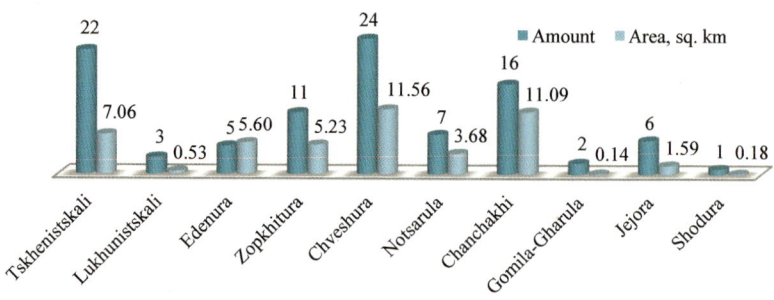

Fig. 3.48 Distribution of the glaciers in the Rioni River basin according to the morphological types

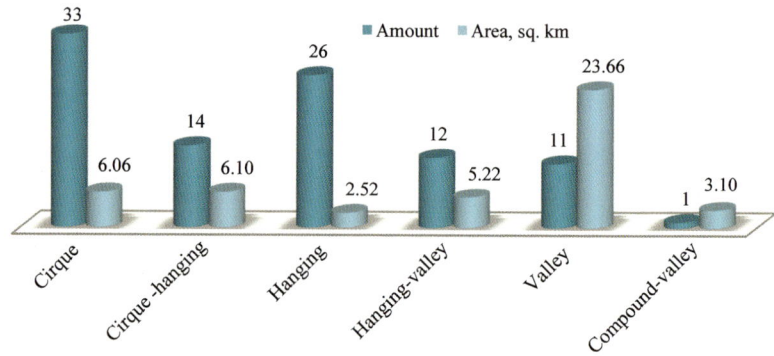

Fig. 3.49 Distribution of the glaciers in the Rioni River basin according to the exposition

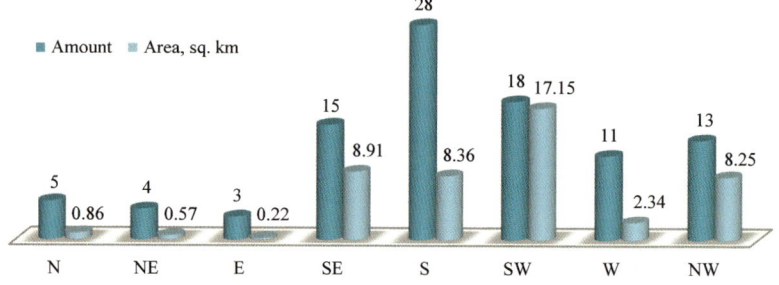

The valley glaciers are located in the southern slopes of the Central Caucasus and the cirque glaciers—mainly in the branch ranges.

It should be noted that there is the only compound-valley glacier, the Buba glacier, in the Rioni River basin, which is located in the Chanchakhi River basin.

The glaciers of the Rioni River basin are mainly located in the southern slopes of the Central Caucasus and Svaneti range. That is why, there dominate the glaciers of the overall southern (S, SE, SW) exposition, both in number and area. Their share in the total number of all the glaciers of the basin is 62.88% and in the total

area of all the glaciers of the basin is 73.76%. The glaciers of the rest exposition are characterized by almost equal distribution (Fig. 3.49).

In order to better characterize the current state of the glaciers of the Rioni River basin, we find it appropriate to discuss it by individual river basins.

There are 22 glaciers with the total area of 7.06 km² in the **Tskhenistskali River basin** by the data of 2014.

Leading place by the number belongs to the cirque glaciers in the Tskhenistskali River basin, and with the amount of the area the valley glaciers are in the first place (Fig. 3.50).

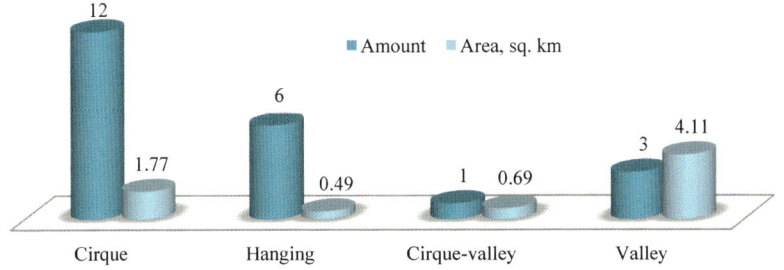

Fig. 3.50 Distribution of the glaciers in the Tskhenistskali River basin according to the morphological types

Exposition of the glaciers is mainly of overall southern. The exposition is determined by the general sublatitudinal direction of the Greater Caucasus and the latitudinal direction of the Svaneti range (Fig. 3.51).

The Tskhenitskali River basin can be divided conditionally into two parts, because these two main centers, the southern slopes of the Central Caucasus and Svaneti range, are distinguished in terms of modern glaciation.

Let us consider these regions separately. 17 glaciers are located on the southern slopes of the Central Caucasus, which includes the two main basins (of the rivers of Zeskho and Tskhenistskali), and their share is 68.18% in the total number and 57.50% in the total area of the glaciers of the Tskhenistskali basin. The glaciers are of mainly cirque and hanging types and have mainly the southern exposition.

The Koruldashi glacier is the largest glacier in the basin. It is the valley glacier of the southern exposition, which is located on the headwater of the river of the same name and encompasses the western section of the Mount Ailama. The glacier's area is 2.20 km².

The glacier is characterized by the open firn valley and by hanging middle section. The glacier tongue is very contaminated by the material brought by the of snow avalanches. In some sections, the loose material reaches a sufficient width, which results in the reduction in the surface ablation in the frontal part of the glacier. There is a vegetation cover in the separate sections of the ice tongue surface. Despite the specificity of the surface of the glacier, the ice tongue reduction is intensive, which is mainly conditioned by the melting of the exposed vertical wall at the end of the ice tongue. Icefall is divided into two ice tongues by the nunatak and flows down to the base of the ledge; the glacier tongue ends at the height of 2680 m above sea level (Fig. 3.52). The firn basin, which is located at the height of 3500 m above sea level, has an oval shape. It is seated in the cirque form and is characterized by a weak inclination.

By the data of K. Podozerskiy the Koruldashi glacier length is 3.2 km, and the area—3.1 km². In the topographic maps of 1960 the glacier area was 3.23 km². The increase in the parameters of the glaciers in this period was caused by the

Fig. 3.51 Distribution of the glaciers in the Tskhenistskali River basin

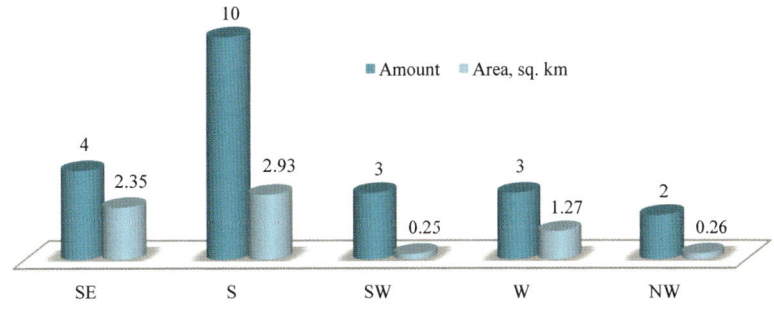

Fig. 3.52 Koruldashi glacier. Google Earth Image, 2013

distorted depicting of its firn valley in the old topographic maps. And in the years of 1960–2014 the glacier area was reduced by 31.88%. Trace of the stade glaciation is well expressed in the Koruldashi River gorge. The lateral moraines make a curve at the end and are transformed into the terminal moraine.

Modern glaciers of the **southern slope of the Svaneti range** are poorly studied. Five glaciers are located there in total; three glaciers—in the Kheledula River basin and two—in the Laska-dura River basin. Their total area is 1.07 km^2. Distribution of the glaciers is related to the central section of the mountain range (as opposed to the glaciers of the northern slope of the range). Mainly small cirque glaciers of southern exposition are located there.

By the data of 1960 there were 23 glaciers on the southern slope of the Svaneti range with a total area of 5.10 km^2, and according to K. Podozerskiy there were only seven glaciers with the area of 3.24 km^2. These data are very small and does not reflect the dynamics of the glaciation. The main reason of it is the incorrect depiction of the glaciers in the old topographic maps. Observation on the latest aerial images shows that in the years of 1960–2014 the glacial cover was changed greatly. The total area of the glaciers was reduced by 79.01%. The largest glacier with the area of 0.68 km^2 is located in the headwater of the Skilori River (the Kheledula River basin).

Lukhunistskali River Basin. Information about the glaciers in this basin is not available in the works of the researchers of early times, which does not allow us to discuss the changes that had happened there. The glaciers of Kareta, Chut-kharo, and Samertskhle are located on the northern slope of the Chudkharo-Samertskhle massif—the highest massif of the Lechkhumi range. By the data of 1960 their area was 1.10 km^2, and by the current situation— 0.53 km^2.

According to the exposition the glacier of Kareta is of northern exposition, and the Chud-kharo and Samertskhle glaciers are of the northeastern exposition. The largest is the Chudkharo glacier with the area of 0.27 km^2.

The Shodura River basin occupies a small area. All of the researchers indicate the only one glacier there, and by the data of 1960 there were two glaciers there with the area of 0.39 km^2. Now there is only one glacier with the area of 0.18 km^2. Shoda glacier is located in the north section of the Shoda peak. Its length is ~800 m and it is of the northwestern exposition. The surface of the glacier is not polluted and it is covered with snow most of the year. By the data of 1960 the second glacier (Shodura) was located in the headwater of the Shodura River and it was the valley glacier. Still today there has been remained a well-expressed moraine of the LIA maximum in the headwater of the Shodura River and the stade moraines of the Late Holocene can be found in the gorge. The third glacier was located on the southeastern slope of the Shoda mountain (in the headwater of the Lagorula River), which morphologically was a small cirque glacier with the area of 0.12 km^2.

By the data of K. Podozerskiy there were four glaciers in the **Edenura River basin**. Two of the glaciers—Didi Edena and Patara Edena occupied the area of 7.25 km^2. By the data of 1960 the number and the area of the glaciers have not been changed. By the data of 2014 there are five glaciers in the Edenura River basin with the total area of 5.60 km^2.

By morphological types there are two cirque glaciers and two cirque-hanging glaciers and one valley glacier—Didi Edena glacier, which is the largest in this basin by the area (Fig. 3.53).

According to the exposition there is one glacier with the southeastern exposition, two—with southern expositions and two more—with the southwestern expositions (Fig. 3.54).

Didi Edena is a valley glacier of the southwestern exposition with the area of 3.52 km^2 (Fig. 3.55). The glacier has a distinct tongue and the feeding basin (firn). Its surface is clean and devoid of the weathered material. Only along the right side stretches the surface moraine, which arises as a result of the weathered material collapsed from the right slope.

At the end of the nineteenth century "the glacier was hanged over the ledge and reached its middle section" (Dinik 1890). Reduction in the ice tongue is caused by the intensive melting on the one hand, and by its mechanical destruction on the other hand. There are the well-expressed moraines of the LIA maximum from the base of the ledge down to the valley. Other morphosculptural forms generated by the glacier are washed.

Zopkhitura River basin. Overall shape of the Zopkhitura River gorge is asymmetrical; the left slope of the gorge is distinguished by soft outlines and a relatively strong erosive fragmentation, than the right slope. Both slopes of the gorge is covered mostly by a quite wide (~5–30 m) dealluvial trails, mounds, debris cones and moraine formations. Cyclic terraces cannot be found in the gorge, however, near the "geologists' flats" and the downstream of the river the well distinct terraced fluvioglacial deposits can be found in the relief. The pluvioglacial terraces, developed around the confluence of the rivers of Zopkhitura and Rioni, are of particular interest.

There are three terraces well expressed in the relief at a height of ~8–12, 25–30, and 40–50 m above the river level. The section of the terraces is presented by the well-processed pluvioglacial cobblestones and sand-sandstone, which is composed mainly of the material of crystalline formations. At the same time, there can be found the large angular granite boulders among these deposits, which may be are the remains of the weakly processed moraines. Such boulders are distributed on the surface of the terraces and in the river bed as well. The moraines of the last glaciation can be found in the same vicinities to the right bank of the Zopkhitura River at the distance of ~500–600 m from the confluence at a height of ~1600–1650 m above sea level.

All of the above facts, together with the other morphological manifestations, provide convincing evidence, that during the Wurm the Zopkhito glacier reached the confluence of the river of the same name, and stretched along the Rioni River gorge, that allows us to restore the boundaries of the last phase glaciation.

Fig. 3.53 Distribution of
the glaciers in the Edenura
River basin according to
the morphological types

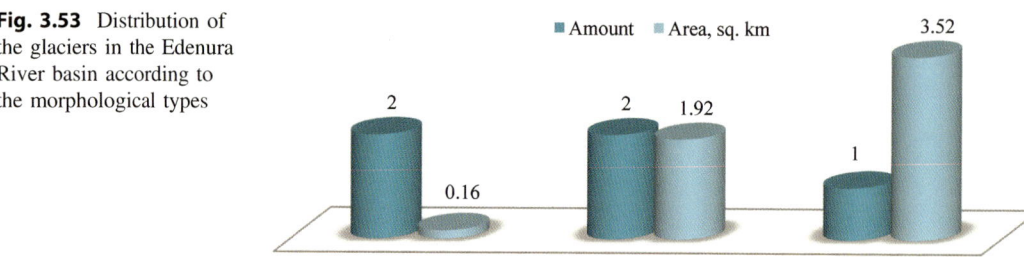

Fig. 3.54 Distribution of the glaciers in the Edenura River basin according to the exposition

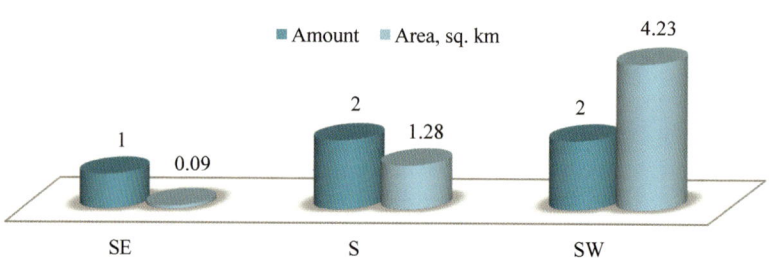

There are well-expressed lateral moraines of the last ice age in the Zopkhitura River gorge at the height of ∼2150–2120 m above sea level, by means of which the restoration of the glacier borders, width, volume, and area is available (Gobejishvili et al. 2012).

Zopkhitura River gorge is of trough shape along to its entire length. By the data of K. Podozerskiy there were seven glaciers in the basin with the area of 7.81 km². By1960 data the number of glaciers was nine with the area of 7.92 km². During this period, the increase in number of the glaciers was caused by their separation during retreating, while the increase in area was caused by a wrong depiction of the cirque glaciers in the old maps. By the data of 2014 there are 11 glaciers with the total area of 5.23 km². During the last 54 years, the area of the glaciers has reduced by 33.96% and their number has increased by 2.

In the Zopkhitura River basin the first place by the amount of the area occupy the valley glaciers; their share is 75.90% in the total area of the glaciers. But according to the number the hanging glaciers occupy the first place (Fig. 3.56).

The glaciers are not equally distributed by the exposition; the glaciers of the southern exposition occupy the first place by the number and by the area dominate the glaciers of the southeastern exposition (Fig. 3.57).

The largest glaciers of the basin are Zopkhito and Laboda glaciers, which were combined until 1955 and ended by one ice tongue; that is why the Laboda glacier is not considered separately in the works of the early researchers. They already were the independent glaciers in the aerophoto images of 1958. Today the ice tongues of the glaciers are about ∼1100 m away from each other. And, the area among them is filled up by the powerful fluvioglacial deposits.

In the years of 2006–2014 we conducted the glacio-geomorphological and glacio-hydro-climatic expeditions to the Zopkhitura River basin. In 2007–2010, we conducted the stationary camp of the Zopkhito glacier for the glacier mass balance research purposes.

The Zopkhito glacier is a simple-valley glacier with southeastern exposition. It is located in the southern slope of the Greater Caucasus and begins to the east of the Geze peak. The

Fig. 3.55 Didi Edena glacier, 2010 (*photo by* L. Tielidze)

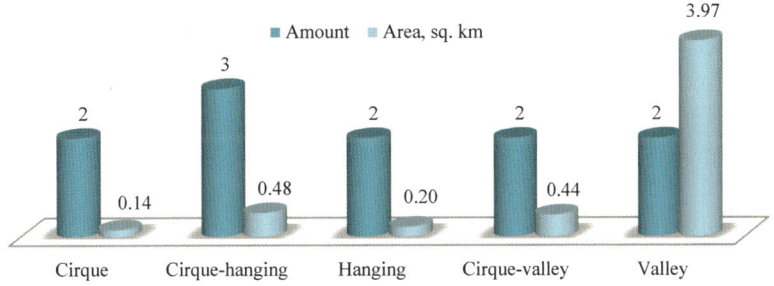

Fig. 3.56 Distribution of the glaciers in the Zopkhitura River basin according to the morphological types

Fig. 3.57 Distribution of
the glaciers in the
Zopkhitura River basin
according to the exposition

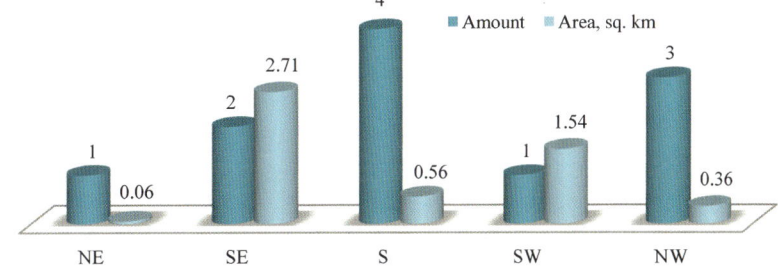

glacier area is 2.42 km^2 and the length—3.60 km by the data of 2014. Its ice tongue ends at a height of 2605 m above sea level. The ice tongue and the glacier's firn basin are separated by the icefall of ∼ 300 m in height; the ice tongue is of the ideal shape and its width is about ∼ 500 m.

Cracks of different direction are developed in the base of the icefall; its surface, especially the sides are covered by the weathered materials (surface moraines) brought by the landslides. The glacier tongue and firn are slightly inclined (∼ 10–20°) and in some places are covered with a moraine cover. Its firn basin is laid up in the sharply expressed cirque, which is located between the peaks of Geze and Edena. On the one hand, it is bounded by the watershed section of the Central Caucasus and on the other hand, by the branch range of the Greater Caucasus, which also is a watershed between the basins of the Didi and Patara Edenas and Zopkhito glaciers.

Like other glaciers in the Greater Caucasus, the Zopkhito glacier intensively retreats as well, which is related to global warming period after the LIA maximum.

Observations on the Zopkhito glacier has begun since the past century. Glacio-climatic (Sh. Inashvili) and glacio-geomorphological (R. Gobejishvili) expeditions were conducted there. The 1:5000 scale maps of the Zopkhito glacier and the neighboring Laboda glacier has been compiled based on the phototheodolite survey (R. Gobejishvili).

The glacier retreating rate within the interval of 1960–1973 has been restored (Gobejishvili 1981, 1989) based on the rows of the moraines in front of the ice tongue (Table 3.3).

The results show that the glacier's total retreat is 61.5 m in 13 years, which an average is 4.7 m a year. But, it should be noted that the indicators of retreat significantly differ according to the years, which is related to the climatic factors. The highest rate of retreat—11–12 m/y was during 1962–1964, while the lowest rate was during 1971–1973. The mentioned phenomenon is related to the 11-year cycle of the solar activity (Gobejishvili 1995).

Some information about the Zopkhito glacier is given below (see Sect. 5.4 Dynamics of the glaciers in 1960–2014, Sect. 5.6 Studying the glaciers dynamics by the micro-stade moraines).

The Chveshura River basin is located on the southern slope of the Greater Caucasus watershed mountain range from the Tsitela peak to the Tsikhvarga peak. There are 24 glaciers there with the total area of 11.56 km^2. By these data the Chveshura River basin is in the first place in the Rioni River basin; its share is 24.74% in the total number and 24.77% in the total area of the glaciers.

The glaciers of Kirtisho, Tsitela, Khvargula, and Shtala are located there. All of the early researchers (Rashevskiy 1904; Podozerskiy 1911; Tsereteli 1959) consider in their works mainly only the glacier of Kirtisho, as it was the largest glacier in the basin. And today the Kirtisho glacier is in the first place by the area not only in the Chveshura, but in the Rioni basin as well.

By K. Podozerskiy there were 17 glaciers in the Chveshura River basin with the area of 20.99 km^2. By the data of 1960 the number of the glaciers increased up to 20, while their total area reduced by 16.27 km^2. Over the 1960–2014

Table 3.3 Zopkhito glacier retreating data of 1960–1973

Years	Retreat, m	Years	Retreat, m
1960–1961	6.7	1967–1968	1.0
1961–1962	3.0	1968–1969	2.0
1962–1963	11.0	1969–1970	2.5
1963–1964	12.3	1970–1971	2.5
1964–1965	7.0	1971–1972	0
1965–1966	6.0	1972–1973	0
1966–1967	5.5		

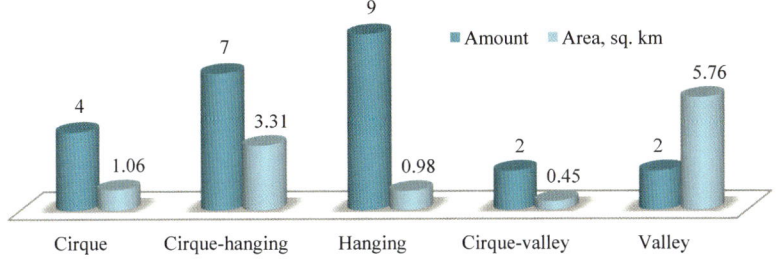

Fig. 3.58 Distribution of the glaciers in the Chveshura River basin according to the morphological types

Fig. 3.59 Distribution of the glaciers in the Chveshura River basin according to the exposition

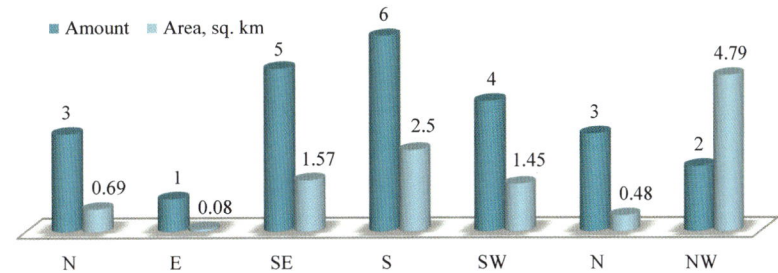

period the number of glaciers increased by 4, and the area reduced by 28.94%.

By the number the hanging glaciers are in the first place in the basin, followed by the cirque-hanging and cirque glaciers. The cirque-valley and valley glaciers have the equal positions, while according to the area the valley glaciers are in the first place (Fig. 3.58).

As the glaciers are located on the southern slopes of the Greater Caucasus, the glaciers of the overall southern exposition dominate there (Fig. 3.59).

The Kirtisho glacier is the largest glacier of the Rioni River basin. By the firn basin it is connected with the firn valley of the Bartui glacier—the glacier of the northern slope of the Greater Caucasus. The width of the feeding basin is ~2.5 km, the length—~2.8 km. The glacier is of valley type with the well-expressed fractured ice tongue. The glacier length is 4.76 km; the width of the middle part of the ice tongue is ~450 m. The glacier surface is clean. Falling of the Chveshura River gorge floor in front of the glacier is very small—~2–3 m/km. Such morphological feature of the gorge leads to the

well-preserved glacial morphosculptural forms: moraines, moraine hills, ram's foreheads, etc.

The glacier's exposition is of northwestern. In the years of 1930–1940 the ice tongue was hanged over the ledge in front of it (Fig. 3.60). Now the ice tongue ends at a height of 2655 m above sea level. Detailed information about the Kirtisho glacier is given below (Sect. 5.5 Valley glaciers reduction after the Little Ice Age maximum).

The Khvargula glacier is the second in size in the Chveshura River basin. It is the valley glacier of the southwestern exposition. In the topographic maps of 1960, it can be seen that the glacier had a well-expressed tongue and glacier area was 1.89 km^2. It should be also said that in the topographic maps of 1960 the firn was outlined incorrectly and a certain part of the feeding basin was placed beyond the Georgia-Russia border, which was caused by the incorrect drawing of the border. Today the various space images clearly show that the boundary line runs through the north mostly, and correspondingly, the glacier firn is completely marked. In spite of the mentioned reason that the firn area was

Fig. 3.60 Ice tongue of the Kirtisho glacier, in 1935 (*photo by* D. Tsereteli)–2010 (*photo by* L. Tielidze)

increased, the area of the glacier was reduced by 29.10% anyway. By the data 2014 the glacier area is 1.34 km².

There are three terraces in the Khvargula River gorge, which are formed during the Late Holocene stade glaciation. The height of the moraine terraces is ∼30–50 m.

Notsarula River basin. By the data of K. Podozerskiy there were six glaciers in the basin with the area of 7.05 km². By the data of 1960, there were six glaciers there with the area of 5.45 km². We note also that in 1960 we were dealing with the six different glaciers. The cirque glacier №357 given in the Podozerskiy's catalog has disappeared, and the glacier of Notsara was divided into two parts; therefore, the number of the glaciers has remained unchanged, while the area was reduced. By the data of 2014 there are seven glaciers in the Notsarula River basin with the total area of 3.68 km². During the last 54 years the number of the glaciers has increased by one, and the area reduced by 32.47%.

Morphologically, the glaciers are not equally distributed; the cirque glaciers occupy the first place by number (Fig. 3.61).

As the glaciers are located on the southern slopes of the Greater Caucasus, their exposition is the overall southern, only there is one glacier with the eastern exposition and another one— with the western exposition (Fig. 3.62).

Fig. 3.61 Distribution of the glaciers in the Notsarula River basin according to the morphological types

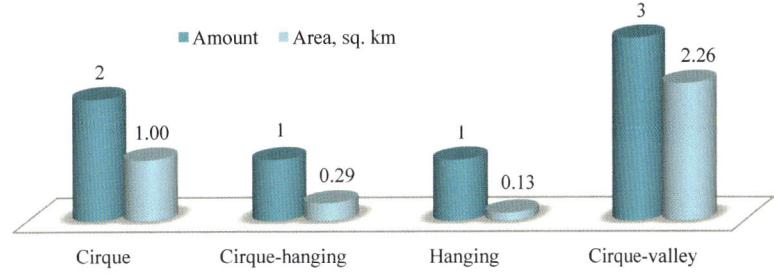

Fig. 3.62 Distribution of the glaciers in the Notsarula River basin according to the exposition

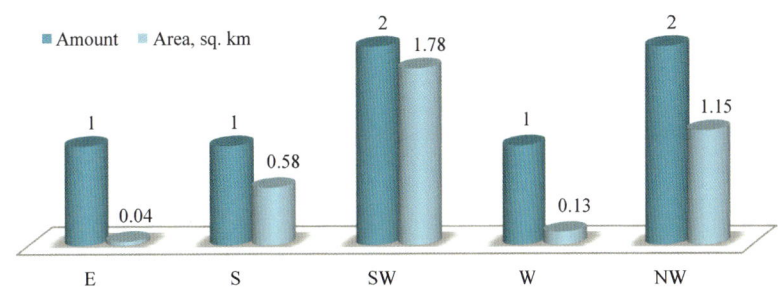

The Kotsoniara glacier is the largest in the Notsarula River basin with the area of 0.96 km². It is a cirque glacier of the southwestern exposition.

The Chanchakhi River basin encompasses the eastern part of the Racha Caucasus and the northern slope of the Shoda-Kedela range. It is the longest tributary of the Rioni River in the mountainous Racha. The highest peaks of the Racha Caucasus are located in this basin with the altitude of over 4000 m (Chanchakhi, Tbilisa, Buba, and Burchula), but the basin is behind the Chveshura River gorge in terms of the area of the glacier, and it has the third position according to the number of glaciers in the Rioni River basin after the Chveshura and Tskhenistskali basins.

Glaciers are found mainly on the southern slopes of the Greater Caucasus (Boko, Buba, Tbilisa, Chanchakhi). But on the northern slope of the Kedela range in the headwater of the Khamijou River only one degraded cirque glacier is remained.

We should note, that almost every year we conducted expeditions in the Chanchakhi River basin in 2005–2014, during which we performed a visual inspection of the glaciers, fixing the dynamics and surveying the ice tongues by the GPS.

There were 15 glaciers with the area of 18.37 km² in the Chanchakhi River basin by K. Podozerskiy. By the data of 1960 topographical maps the area of the glaciers were reduced to 14 and the area—by 3.57 km². We also note that the reduction in the number of the glaciers in this period was caused by the disappearance of the small glaciers on the northern slope of the Shoda-Kedela mountain range. The decrease would be higher, if the small glaciers would not have separated from the Tbilisa glacier. By the data of 2014 there are 16 glaciers in this basin with a total area of 11.09 km².

The valley glaciers are in the first place by the area in the Chanchakhi River basin, while by the number of hanging glaciers occupy the first position, but their area is the smallest (Fig. 3.63). We should note that the Buba glacier—the compound-valley glacier is located in the Chanchakhi River basin, which is the only compound-valley glacier in the Rioni River Basin.

The distribution of the main glaciers on the southern slope of the Greater Caucasus determines their exposition as well; therefore, the glaciers of the southern exposition occupy the first position by the number and area (Fig. 3.64).

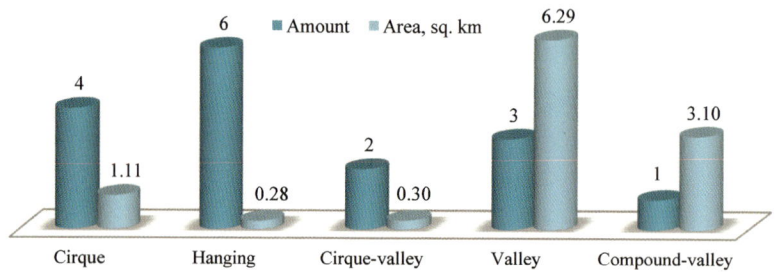

Fig. 3.63 Distribution of the glaciers in the Chanchakhi River basin according to the morphological types

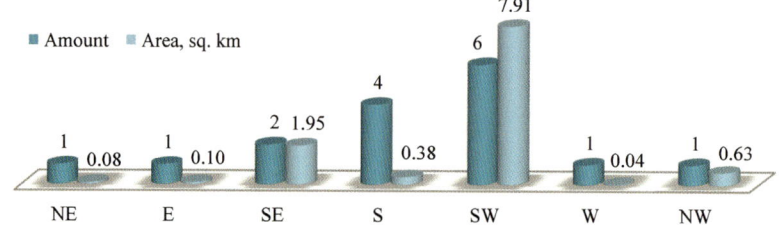

Fig. 3.64 Distribution of the glaciers in the Chanchakhi River basin according to the exposition

The Boko glacier is in the first place in the Chanchakhi River basin by its morphometric parameters. Its area is 3.88 km^2. It is a valley glacier of the southwestern exposition and after Kirtisho it occupies the second place in the Rioni River basin by the area.

The Glacier firn and ice tongue are well separated from each other, the glacier tongue is laid up for about 2 kilometers in the valley. Its width is ∼400 m and is fractured by the cracks of different directions. The firn of the Boko glacier is stretched at a distance of ∼3.8 km from the northwestern to the southeastern direction. The firn is connected to the Karaugom glacier of the northern slope of the Greater Caucasus. A width of the firn at the place of junction is ∼1.0 km. The ice tongue surface is mostly clean.

In 1960, the glacier area was 4.62 km^2 and the length—4.8 km. The glacier tongue was ended at a height of 2440 m above sea level. By the data of 2014 the glacier area is 3.88 km^2 and the length—4.10 km. The Boko glacier tongue ends at a height of 2660 m above sea level. During the last 54 years the glacier area was reduced by 16.01% and the ice tongue retreated by ∼700 m, but it rose up by ∼120 m.

It is possible to identify the drastic change of the Boko glacier by comparing of the old and modern images each other (Fig. 3.65).

The trace of the last stade glaciation is well expressed in the Bokostskali River gorge. The lateral stade moraines stretch along its both sides, which make curve below and transform into the terminal moraine. Outer slopes of the lateral moraines are slightly inclined and are separated from the main slope by a depression, where there is a small lake there. The inside slope is very inclined and fragmented by erosion and gravitation processes. Micro-stade moraines can be well seen in the base of the valley.

The Buba glacier is a compound-valley glacier of the southwestern exposition (Fig. 3.66). It consists of three flows and with the combined broad ice tongue ends at 3020 m above sea level in the valley. A detailed description of the glacier is given below (Sect. 5.4 Dynamics of the glaciers in 1960–1985). We would just like to say that the glacier is called Tbilisa by a mistake; in all the early editions and what is important, by a local population it is named as Buba. Over the years, the Vakhushti Bagrationi Institute of Geography conducted stationary studies

Fig. 3.65 Boko glacier reduction in the years of 1970 (*photo by* R. Gobejishvili)–2008 (*photo by* L. Tielidze)

Fig. 3.66 Buba glacier, 2012 (*photo by* L. Tielidze)

of glacio-geomorphological and glacio-hydro-climatic direction of the glacier.

It should be noted that during the period of the Soviet Union there was the longest series of stationary observations on the Buba glacier (Tbilisa) from the year of 1965, which was conducted by the Institute of Geography. These studies were terminated in the 90s of the last century.

The Tbilisa glacier is located in the Tbilisa basin—the left tributary of the Bubistskali River. It is a valley glacier of southeastern exposition. Last time we surveyed the ice tongue by using the GPS in 2010; that time the last section of the tongue was hanged over the ledge. During the expedition of 2012, the surveying of the glacier tongue by GPS was impossible, because due to the mechanical destruction and ablation the last section of the tongue was collapsed; accordingly, due to the high elevation of the ledge we could not even approach it. Therefore, it can be said that the surveying of the dynamics of the glacier will be available only by aerial images, and photo shooting (Fig. 3.67).

By the data of 2014 the glacier area is 1.90 km². In this regard, Tbilisa is the third glacier after the Boko and Buba glaciers in the Chanchakhi River basin. Its tongue ends at a height of 2990 m above sea level. The glacier surface is cracked. Its length is 2.87 km.

By the data of 1960, the Tbilisa glacier area was 2.76 km² and its tongue was ended at a height of 2820 m above sea level. During the last 54 years the area of the glacier was reduced by 31.15%. The ice tongue was raised by 170 m.

There are the well-expressed stade and micro-stade moraines in the glacier basin, which allow reproducing the dynamics of the glacier in time and space.

There are only two small cirque glaciers in the **Gharula and Gomila River basins** with the area of 0.14 km². The glaciers are located on the southern slope of the Shoda-Kedela range. During the last 54 years they underwent the significant changes both in number and area. By the data of 1960 there were five glaciers with the area of 0.82 km².

There are modern glaciers in the **Jejora River basin** on the southern slope of the Greater Caucasus, to the south of the Mount Saukhokhi and on the northwestern slope of the Leti range. According to the recent data there are six glaciers in total, four of them of cirque type and two—of cirque-valley type with the total area of 1.59 km².

Fig. 3.67   Tbilisa glacier reduction, in the years of 2010 (*photo by* L. Tielidze)–2012 (*photo by* L. Tielidze)

Fig. 3.68   Distribution of the glaciers in the Jejora River basin according to the exposition

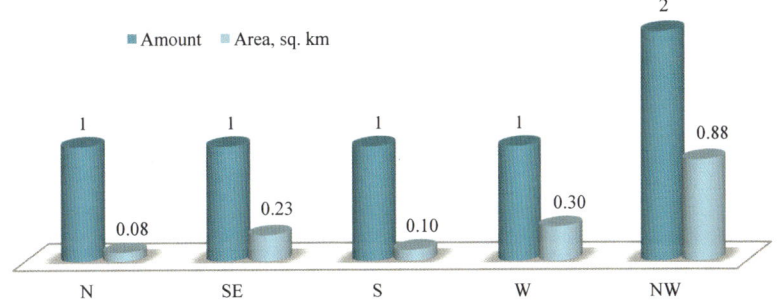

According to the exposition only two glaciers are of northwestern direction, while the rest of the glaciers are of different expositions (Fig. 3.68).

By K. Podozerskiy there were 11 glaciers with the area of 7.24 km^2. By the data of 1960 there were nine glaciers in the basin with the total area of 5.73 km^2. From the year of 1890 the Zekara glacier was divided into two parts and its area was decreased by 0.7 km^2. After the year of 1960, the southern Zekara was divided into two parts again. Khalatsa glacier located in its both sides was melted at all together with the three small cirque glaciers. During the last 54 years, the area of the glaciers was reduced by 72.25% in this basin.

3.2.7   Glaciers of the Liakhvi River Basin

The Brutsabdzeli (Java) range (the branch-range of the Greater Caucasus) and the Tergi and Liakhvi River basins' watershed Caucasus range nearby the Mount Zilgakhokhi and Mount Laghztsiti are the main centers for the modern glaciation in the Liakhvi River basin. These sections are the highest hypsometrically in the research area.

By the data of K. Podozerskiy there were 12 glaciers with the total area of 5.15 ± 0.13 km^2 in this basin. By P. Ivankov there were 34 glaciers with the total area of 10.41 km^2. To our opinion, his data are increased and do not match reality.

Fig. 3.69 Distribution of the glaciers in the Liakhvi River basin according to the exposition

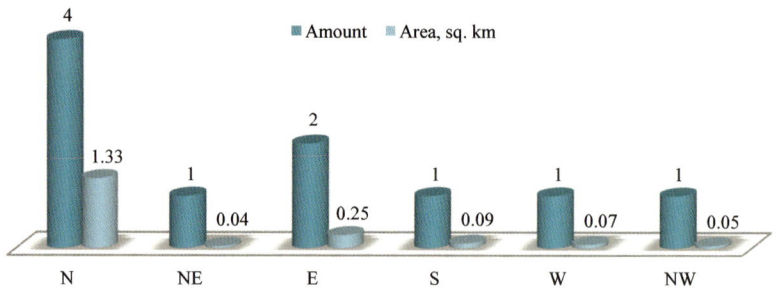

It should be noted that P. Ivankov considered the seasonal snow spots as glaciers, by which this basin is very rich. According to the data of 1960, there were 16 glaciers in total in the Liakhvi River basin with the total area of 4.27 ± 0.13 km^2. The data of K. Podozerskiy and 1960 well reflect the dynamics of the glaciation in this basin and agree with those times dynamics of glaciation in the Caucasus in general.

By the data of 2014 there are 10 glaciers in the Liakhvi River basin with the total area of 1.82 ± 0.07 km^2. Based on the above mentioned, we can say that the glaciers have reduced in the area by $22.24 \pm 2.78\%$ during the period of 1911–1960, but the number of glaciers has increased by four during the same period. And during the last 54 years the glaciers number has decreased and their area has reduced as well.

The Laghztsiti glacier is the largest glacier in the Liakhvi River basin. It is the only glacier of the cirque-valley type of the northern exposition in this basin. In the plan of the aerial image it is seen that the glacier has the two tongues which are very far to each other, but in fact the eastern ice flow is a continuation of the firn valley, and the western—is the real tongue of the glacier; difference in elevation is ~ 300 m. Glacier area is 0.99 km^2 and its share in the total area of glaciers in the Liakhvi River basin is 54.09%. It is located in the Ermanistskali River basin.

The rest of the area is distributed among the small size of glaciers of cirque type, one of which is located in the southwestern slope of the Greater Caucasus to the south-west of the Mount Zilgakhokhi and another glacier is located in the place, where the Leti mountain range is separated from the Greater Caucasus. The rest of the seven

glaciers are located on the slopes of the Java range. According to the number the glaciers of the northern exposition dominate, while the glaciers of the rest of the expositions are nearly evenly distributed (Fig. 3.69).

The glaciers are mainly located in the basins of the Chomaghistskali, Keshelta, and Ermanistskali Rivers.

3.2.8   Glaciers of the Aragvi River Basin

Unlike the Central Caucasus, the Aragvi River basin is characterized by relatively low hypsometric location characteristic peculiarities of weather conditions added to it; in particular, the impact of the dry continental air masses. As a result of above-mentioned, today only one glacier of Abudelauri is presented in the Aragvi River basin (Khevsureti's Aragvi) with the area of 0.31 ± 0.015 km^2. By the data of 1960 there were two more glaciers in the Aragvi River basin except the Abudelauri glacier, which were located in the Keli volcanic plateau in the Tetri Aragvi River basin. Today these glaciers do not exist anymore. The total area of those three glaciers was 0.88 ± 0.03 km^2. By the data of K. Podozerskiy the total area of the three glaciers was 2.21 ± 0.04 km^2.

The Abudelauri glacier is a cirque-valley glacier of the northeastern exposition (Fig. 3.70). In 1960, the glacier area was 0.68 km^2 and its length—2.19 km. By the data of 2014 the glacier area is 0.32 km^2 and the length—1.36 km.

The glacier is deep-seated in the old glacial cirque. The ice tongue is polluted and the modern

Fig. 3.70 Abudelauri glacier, 2004 (*photo by* R. Gobejishvili)

stade morphosculptural forms are weakly expressed. However, there are well-preserved Late Pleistocene and Holocene age moraines in the base of the Roshka gorge, which enable the reconstruction of the old glaciations on the southern slope of the Eastern Caucasus.

3.2.9 Glaciers of the Tergi River Basin

Headwater of the Tergi River basin is located to the north of the Greater Caucasus in Georgia. It is bordered to the south by the main watershed range peaks—Vatsisparsi, Kalasani, Laghztsiti, Laghatisari, Khorisari, Narvani, etc.; to the south-east and north-east—by the ranges of the Greater Caucasus and Shavana; to the north and north-west—by the Khokhi and Chachi ranges.

The Tergi River flows along the Truso depression from the headwater to the site of Kasriskeli. The gorge has a submeridional direction from Kobi to Kazbegi, and meridional—after Kazbegi. It includes Khevi—a historical-ethnographic region of Georgia. The beauty of Khevi is the Mount Mkinvartsveri (Kazbegi), 5033 m in height, one of the highest peaks of the

covered with the snow-ice coat Greater Caucasus and Europe (Fig. 3.71).

It should be noted that during 2005–2014 we were intensively conducting the expeditions almost to all glacier river basins in the Tergi River basin annually. During the glaciers monitoring process we surveyed the ice tongues by means of GPS, also conducted their marking, identification of the dynamics and visual observation.

The main center of the glaciations in the Tergi River basin is the Jimara-Kazbegi massif (Khokhi range). The average height of this massif is ~4500 m. Powerful valley glaciers, such as: Devdoraki, Gergeti (Ortsveri), Mna, Suatisi, and others come from the slopes of this massif. Individual centers of the glaciations are connected with the peaks of the main watershed range of the Greater Caucasus the height of which is 3800 m (Zilgakhokhi, Kalasani) and with the Khde River basin (Gobejishvili et al. 2013) (Fig. 3.72).

According to individual river basins the glaciers and the areas occupied by them are unequally distributed (Fig. 3.73).

By the data of K. Podozerskiy there were 63 glaciers in the Tergi River basin with the total

Fig. 3.71 Mount Mkinvartsveri (Kazbegi), 2010 (*photo by* L. Tielidze)

Fig. 3.72 Distribution of the modern glaciers of the Tergi River basin according to its tributary river basins

Fig. 3.73 Distribution of the glaciers of the Tergi River basin according to its tributary river basins

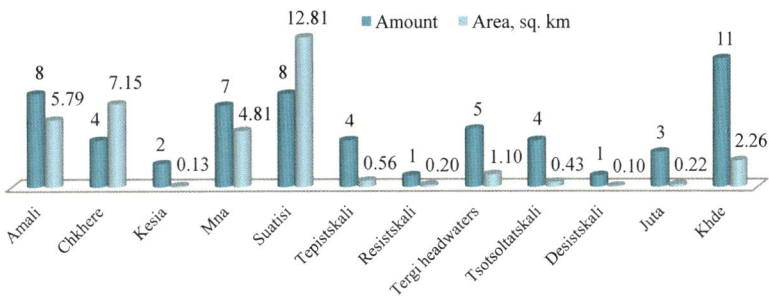

area of 89.12 ± 1.22 km². By the data of the topographical maps of 1960 there were 99 glaciers in the Tergi River basin with the total area of 67.01 ± 1.33 km².

There are 58 glaciers by the data of 2014 with the total area of 35.56 ± 0.8 km² in the Tergi River basin. According to the number of glaciers the Tergi River basin is in the fourth place after Enguri, Kodori, and Rioni and its share is 9.10% in the total number of glaciers of Georgia. It is also in the fourth place by the area after Enguri, Rioni, and Kodori and its share in the total area of the glaciers of Georgia is 9.99%.

During the last 54 years, the number of glaciers has decreased by $\sim 41.41\%$, while the area has reduced by $47.07 \pm 2.12\%$. By these data of reduction in number and area of the glaciers the Tergi River basin is leading among the four bigges river basins (Enguri, Rioni, Kodori, Tergi) of Georgia. The reason of it is the several natural factors, the following three key factors of which we singled out:

1. As we mentioned above, the Tergi River basin is located to the north of the main watershed range of the Greater Caucasus and the climate is sharply continental, which causes a lack of solid atmospheric precipitation (this is particularly noticeable in recent years), which is one of the necessary conditions for the existence of glaciers and their feeding regime;
2. Orographical conditions are also important; in particular, the relief's fragmentation is quite high, and the height of several peaks exceeds

4000–5000 m, and there are abundant peaks in these places with the height more than 3500 m; due to this the slope inclination is high, which directly affects the morphological types of glaciers. Compared to the other river basins of the Georgia, there is a biggest amount of hanging glaciers in percentage in the Tergi River basin, the share of which is 41.37% in the total number of the glaciers in this basin. If we add the glaciers of cirque-hanging (8.62%) and hanging-valley (13.79%) types, it means that 63.78% is the share of the glaciers of the overall hanging morphological types (cirque-hanging, hanging-valley, hanging) (Fig. 3.74).

Due to such conditions, the mechanical destruction of the glaciers is frequent in this region, which hastens the ice melting intensity. In addition, the mechanical destruction of the glaciers is one of the key contributing factors to glacial torrents. The mentioned glacial torrents in many cases lead to the collapse of infrastructure in the mountainous regions, destruction of the roads and bridges, and in some cases, human victims. Such cases are common in Georgia. Lately, in May 17, 2014 such torrent was recorded in the glacier of Devdoraki, which claimed the lives of several people. A detailed discussion of this process we have given below, describing Devdoraki glacier:

3. Furthermore, in recent decades, the increase in temperature is observed as in the Tergi River Basin as well as in many mountain regions of the world, which hastens the

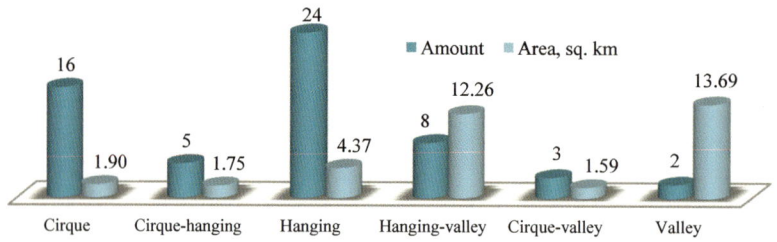

Fig. 3.74 Distribution of the glaciers in the Tergi River basin according to the morphological types

melting process of glaciers. In places where the glaciers are often small in size, their number decreases in parallel to the reduction in their area. That is the case of the 1960–2014, when such a high rate of the reduction in number and area of the glaciers was recorded in the Tergi River.

According to the area the leading place, occupy the valley glaciers in the Tergi River basin, the share of which is 38.49% in the total area of the glaciers. The hanging-valley glaciers are in the second place—34.47%, and then come the glaciers of the rest of the morphological types. Cirque-valley glaciers occupy the smallest area. As it was mentioned above, according to the number the hanging glaciers are in the first place—41.37%, and there are only two glaciers of valley type (Fig. 3.74).

Glaciers of all expositions can be found in the Tergi River basin. According to the number, the glaciers of the northern exposition occupy the first place; then follow the glaciers of the north-western exposition. Glaciers of the southeastern and southern expositions are equally distributed, as well as the glaciers of the southwestern and western expositions. Of these, the latter is in the last place according to the area. And, the glaciers of the southwestern exposition occupy the first place according to the area. In total, according to the area the glaciers of overall southern exposition are in the first place, and according to the number, the glaciers of the overall northern exposition are in the leading place (Fig. 3.75).

The Tergi River basin glaciers are well studied in different years by different researchers, so we stop at a brief description of the individual basins.

The Amali River basin is located on the northeastern slope of the Jimara-Kazbegi massif. It is a left tributary of the Tergi River. There are eight glaciers with the total area of 5.79 km^2 by the data of 2014. And by the data of 1960 there wereseven glaciers and their total area was 11.74 km^2. During the last 54 years the glaciers in this basin were modified greatly. In particular, the three small glaciers have melted completely on the southeastern slopes of the Chachi glacier; the Chahi glacier area has been substantially reduced and divided into two parts; one unnamed glacier between Devdoraki and Chachi glaciers was almost halved and split into three parts; the small-size glacier to the east of the southern Devdoraki was melted completely and does not exist anymore; as for the Devdoraki glacier itself, it was divided into three parts, namely into the two main and one small-size glaciers. Thus, within the period of 1960–2014 the glaciers area was reduced by 50.68% in the Amali River basin, while their number was increased by one.

As for the morphological types of glaciers, all of the glaciers are of hanging type, except for the northern Devdoraki and southern Devdoraki, which are the hanging-valley glaciers.

Orographical peculiarities of the relief condition the overall northern exposition of the glaciers there (glaciers of other exposition cannot be found in this basin); out of which the glaciers of northeastern exposition occupy the leading position. There is only one glacier of north-western explosion and another one—of northern exposition (Fig. 3.76).

The Northern Devdoraki and Southern Devdoraki glaciers with a total area of 4.0 km^2 are the largest glaciers in the Amali River basin.

Fig. 3.75 Distribution of the glaciers in the Tergi River basin according to the exposition

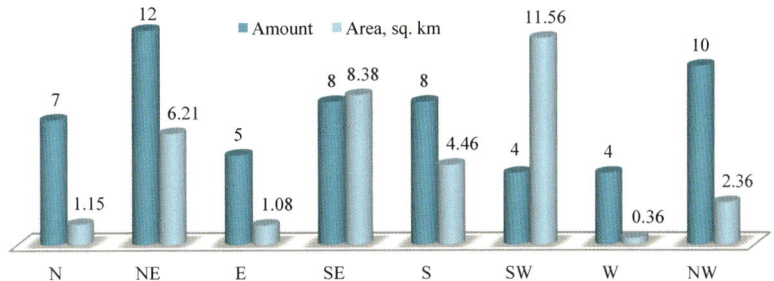

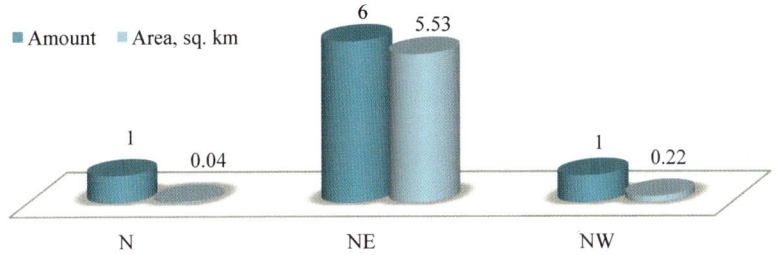

Fig. 3.76 Distribution of the glaciers in the Amali River basin according to the exposition

Both of them are the hanging-valley glaciers with the northeastern exposition. In 1960, the mentioned glaciers were represented in the form of one glacier and its area was 4.82 km².

As mentioned above, the Devdoraki glacier was broken up into three parts, hypsometrically higher southern flow was separated from the northern flow. Aerial images show, that the glaciers should have been split around the year of 2000. The third small hanging glacier is located to the southeast from them at a height of ~4000 m above sea level.

During the last 54 years the glacier area has reduced by 17.01%; three glaciers were formed in its place. Greater part of the northern Devdoraki and southern Devdoraki is located beyond the Georgia's boundary.

We have obtained the old topographic map of the Mkinvartsveri massif, which was drawn up in 1882, which clearly shows that the Devdoraki glacier flowed to much lower elevation and accordingly, the area was much more than in 1960. In addition, we looked after the Devdoraki glacier photos in the TSU Geographical Museum, taken in 1880–1897, approximately in

the same period. We compared the image with the photo of the modern glacier taken from the same place in 2011; It is seen even visually, that the reduction is obviously high (Fig. 3.77).

Devdoraki glacier is known for its powerful ice avalanches, which were mentioned in 1776, 1778, 1785, 1808, 1817, and 2007. In 1832, there was a powerful ice avalanche, when the icy avalanche blocked the Tergi River and stopped its flow for 8 h (Gobejishvili 1995). After breaking the dam the glacial torrent made a great damage to Vladikavkaz city.

Last similar natural disaster occurred in Georgia in May 17 of 2014 at 10 a.m. When in the Dariali gorge (Georgian Caucasus, Kazbegi region), in the confluence of the Tergi and Amali (Devdoraki) Rivers the action of catastrophic rock-ice avalanche and glacial mudflow took place, which caused the full paralyze of the functionality of the infrastructural objects of strategic importance for the country. Disaster damaged the 134–135 km section of the Mtskheta-Stepantsminda-Larsi highway (Georgian military road) of the International importance (in total, about 1600 m of the road section).

Fig. 3.77 Reduction of the Devdoraki glacier, 1880 (**a**) (*photo by* Museum of Geography at TSU)—1897 (**b**) (*photo by* Museum of Geography at TSU)—2011 (**c**) (*photo by* L. Tielidze)

The traffic was prohibited from May 17 to June 14. There worked 70 units of road engineering. International gas pipeline (Trans Caucasus gas pipeline) and under construction hydropower were damaged. Mudflow damaged the high-voltage power transmission tower, the border guard base was isolated from the outer world, as well as the customs checkpoint and residence of the Patriarchate of Georgia, and created everyday problems to their personnel (Fig. 3.78). The disaster resulted in the death of nine people.

ASTER satellite image taken one day before the disaster in cloud-free weather condition in 16 May of 2014 and Landsat L8 OLI image taken approximately two months after the disaster in a little cloudy weather condition in 11 July of 2014 we used for investigation. Despite of clouds it is quite easy to distinguish results of the disaster.

On the ASTER image taken in 16 May, 2014 it is shown that starting zone of rock-ice avalanche was unstable even before, because trace of little rock fall is observed (Fig. 3.79a). The same is noticeable on the Landsat images taken in 2009 (comparatively little scale rock fall). Accordingly it can be assumed that one of the reasons (together with other reasons) for calving of one rock-ice mass and developing of disaster were a small rock falls commenced in past years.

The result of the disaster is well shown on the Landsat image taken in 11 July of 2014 (Fig. 3.79b). The mentioned data and aerial images taken from the helicopter give us opportunity to reconstruct rock-ice avalanche and glacial mudflow.

Unlike the phenomena occurred in the gorge in the past, when the ice masses were extracted from the end of the glacier tongue, the rock-ice avalanche forming zone of 2014 is located at relatively high benchmarks, namely, at a height of 4500 m on the very steeply inclined northeast slope of the Kazbegi peak (Fig. 3.80a, b). In the rightmost part of the Devdoraki glacier's feeding zone occurred the rock-avalanche of the rocky part and glacier part located in it and the grand scale snow masses. Collapsed mass was transformed into the rock-ice avalanche, which was divided into the three flows and sprinted down the steeply angled ice tongue surface with high speed (NEA Report 2014) (Fig. 3.81).

One of the flows jumped over the rocky tip spraying the transformed mass to the steeply angled slope, but the most part of the material collapsed instantly in the glacier again. Reddish-pink trace is observed vividly in the high benchmarks of the left slope in the form of a large spot (Fig. 3.82a). The largest left flow placed the central part of the ice tongue in the dynamical zone and the mentioned part of the ice tongue suffered the strong deformation and fragmentation into separate ice blocks (NEA Report 2014).

Fig. 3.78 Results of the Devdoraki glacier's glacial mudflow in the Dariali gorge, May 18, 2014 (Aerial image PLEIADES 18.05.2014)

Below this section all flows were united in a single flow of grandiose volume and rushed down to the gorge at a high speed and in its way took away the glacier deposits and weathered erosion material allocated in the Devdoraki River bed and in its slopes. Mudflow transported high-structural material sank into the Devdoraki River bed of variable width (15–30 m). The flow, which went out of the narrow part of the sharply bended section before the Amali River confluence at a height of 30–35 m washed the right slope's fragment (trace of such washing has not been observed yet in the morphology created by past mudflows at relative height of 55–60 m

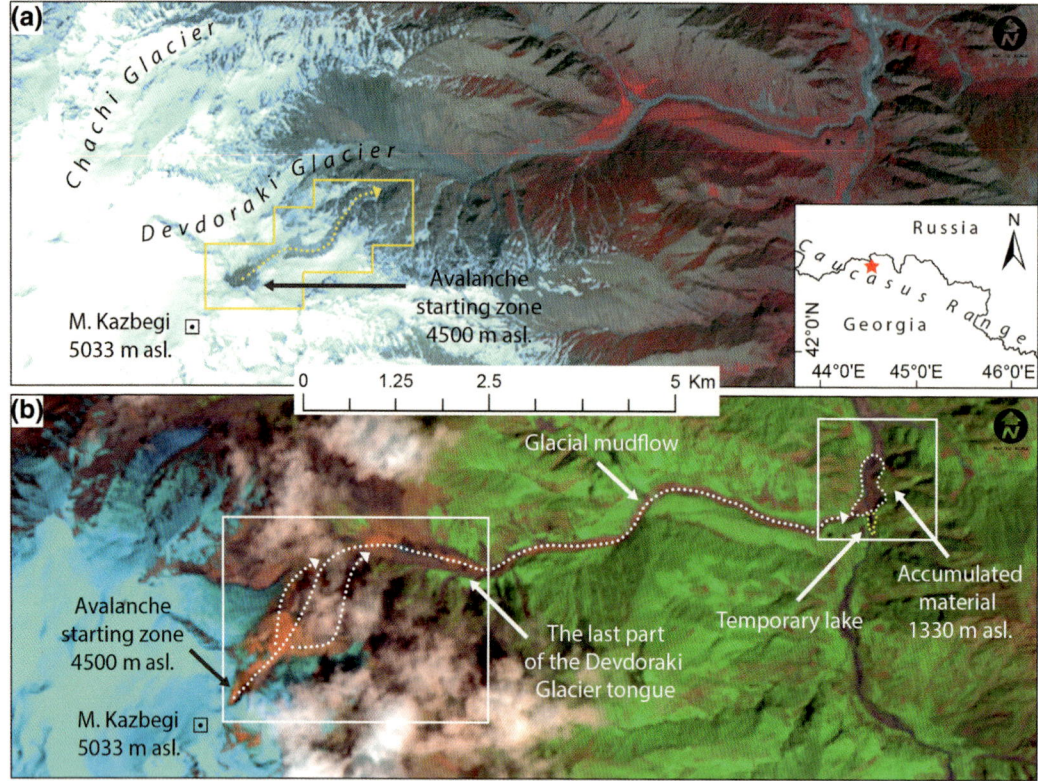

Fig. 3.79 **a** ASTER satellite image taken in 16 May, 2014. **b** Landsat L8 satellite image taken in 11 July, 2014

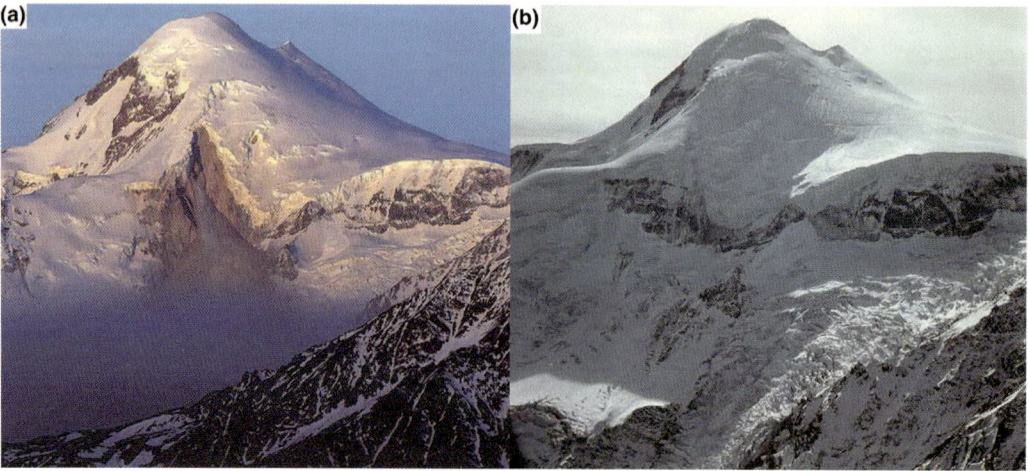

Fig. 3.80 Mt. Kazbegi—5033 m a.s.l. **a** The hearth of Rock-ice avalanche genesis, 17 May of 2014 (*photo by K. P. Rototaeva*). **b** The same site taken in 1987 (*photo by K.P. Rototaeva*) (Kotlyakov et al. 2014)

from the bed). Below the turn, both the gorge and its bed gradually extend to the Amali River conically. In the mentioned section of the river bed the boulders of large size (3–6 m) and rolling boulders of moraine genesis were deposited, which can be explained by the extension of the

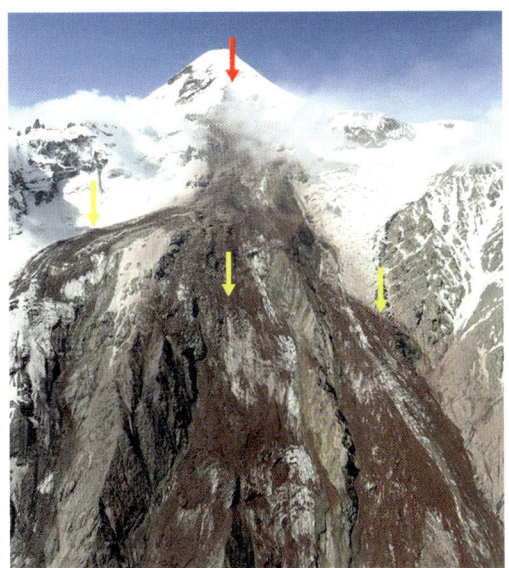

Fig. 3.81 Photo taken from the helicopter in 17 May of 2014 (*photo by* G. Gotsiridze). *Red arrow* shows the center of genesis of the rock-ice avalanche. *Yellow arrow* shows the three flow of the rock-ice avalanche

mentioned part of the gorge after the sharp bend, by reduction of the kinetic energy of the flow and shifting the motion vector in the direction of the left slope. Therefore, the trace of the wavy passage of the flow on the slope is not uniform and varies within 3–15 m. After joining the rivers of Devdoraki and Amali the flow was moving in the curved bed of the trapezoidal cross-section developed in the fluvioglacial mudflow sediments. Cutting depth is about 20–25 m in average. From the place of joining of the rivers of Devdoraki and Amali up to the Gveleti Lake meridian, the mudflows passage traces in the curved bed slopes are observed at the different attitudes and ranges within 10–20 m from the bottom of the valley. It is notable that in the material deposited on the slopes of the gorge due to gradual ice melting, the melted forms are formed and relatively large fraction material fell into its bed. In the Gveleti Lake meridian the river bed turns sharply to the northeast. In

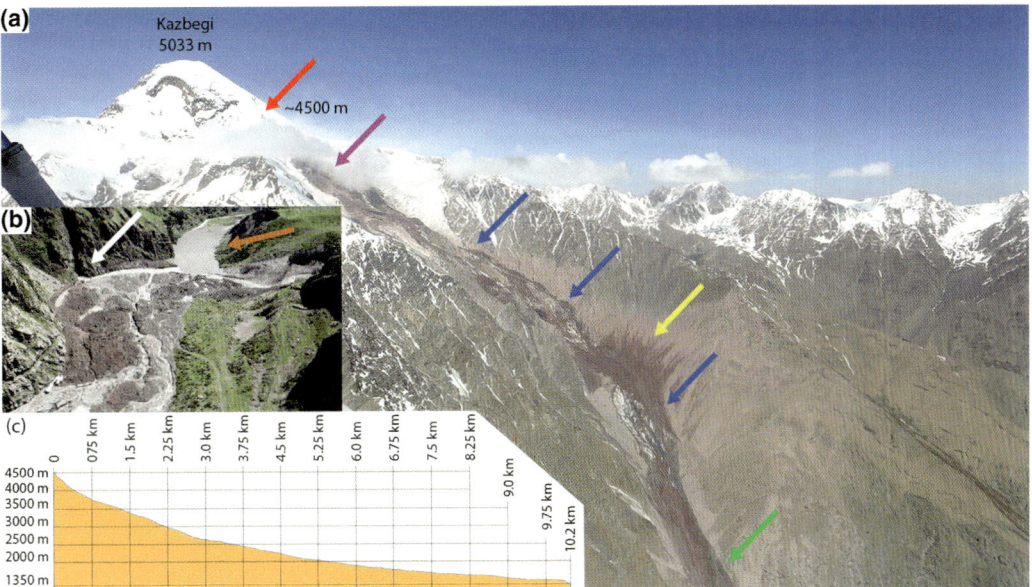

Fig. 3.82 **a** Photograph taken from the helicopter of 20 May, 2014 (*photo by* G. Gotsiridze): With *red arrow* is shown cut spot of Rock-ice avalanche; *yellow arrow* shows the spraying place of the transformed material; *lilac arrow* shows accumulated material in the separation cirque areal; *blue arrows* shows accumulated material in the Devdoraki ice tongue areal; *green arrow* shows accumulated material in the transit zones of the gorge. **b** Tergi gorge, the result of glacial mudflow: *Brown arrow* shows the temporary lake, *white arrow* shows the "DarialHES" transport tunnel from which the accumulated water flow discharged. **c** Hypsometric profile from breaking Rock-ice avalanche up to final stop

addition, the trapezoidal section's relatively wide bed transforms instantly into a narrow "V"-shaped cross-section bed. Accordingly, its throughput is considerably limited. Therefore, in the place of sharp bend the sort-term uneven temporary damming of the mudflow took place, due to which the flow wave deposited the mudflow material in the edge of the slope, the height of which is 25–30 m compared with the left slope. Below this section the cutting depth is up to 23–25 m; due to the temporary damming of the flow, its pressing energy increased and the flow was thrown to the direction of the confluence of the Tergi River, reached the rocky cliff composed of granitoids and deposited the solid material transported by the mudflow at its foot. In addition, the part of the flow began flowing to the northward and stopped about 100 m away from the hydro technical building of the DarialHES (NEA Report 2014).

As a result of accumulation of solid sediment transported by mudflow, a powerful debris cone has been occurred (of about 1.5 mln m^3 in volume), which completely blocked the Dariali gorge (Fig. 3.82b). From the wall of separation up to the Tergi River confluence the overall volume of the withdrawn mudflow mass made about 5–6 mln m^3, the most part of which was accumulated in the area of the separated cirque and in the Devdoraki ice tongue and transit zones of the gorge (Fig. 3.82a).

Overall, the Devdoraki glacial debris rock-mudflow was characterized by high density. Without high density and corresponding geological nature, it would be impossible to bring such large boulders to the erosion basis. Moreover, the mudflow was characterized by such a high density that even after his entry into the main river bed the Tergi River could not provide the liquefaction of the rock-mudflow and transferring to the downstream of the river. It should be noted that in the first stage of the accumulation of debris flow in the Tergi River gorge, the important positive role played the transport tunnel of "DarialHES" in the passage of the certain part of dammed water flows (NEA Report 2014) (Fig. 3.82b). If there was not the water discharge

from the temporary lake through a transport tunnel, we would face the more catastrophic consequences.

The mudflow after the coming out into the Dariali (Tergi) gorge has blocked the Tergi river for some hours and approximately 20–30 m depth lake was created (Figs. 3.78 and 3.82b), the water volume made at least 150,000 m^3. The falling from the hearth of Rock-ice breaking to Dariali gorge (place of accumulated material) is ∼3.2 km., while distance is ∼10.2 km (Fig. 3.82c). The disaster had a devastating effect for such big inclination and distance. According to preliminary estimates V.N. Drobyshev, the maximum speed of the flow could be 280 ± 30 km/h (Chernomorets 2014).

According to the research we can conclude, that activation of glacial mudflows and rock-ice avalanches in the Tergi River basin along with the high energetic potential of the relief is stipulated by the following reasons:

1. Active movement of morpostructural blocks separated by tectonic faults, high seismic risk, and the critical tension of physical fields. During the last 100 years, in the Kazbegi seismic detection block the earthquakes of 7–8 magnitude intensity were observed in 1878, 1915, 1947, 1951, and 1992 (NEA Report 2014);

2. Different relation of the volcanic and sedimentary rocks to the weathering agents (NEA Report 2014);

3. Existence of deposit of large volume of the glacial, fluvioglacial, and moraine formations in the gorge beds and their slopes (NEA Report 2014);

4. By the percentage the most general hanging (circus-hanging, hanging-valley, and hanging) glaciers are located in the Tergi river basin and they make 63.8% of the whole glaciers of the Tergi River basin. The mechanical breaking of glaciers is frequent in the region, which accelerates intensity of melting. Also it is one of the supporting factors for rock-ice avalanches and glacial mudflows;

5. The long-term average air temperature of 1961–2009 on Jvari pass and High Mountainous Kazbegi weather stations (same region) is higher by 0.3 and 0.2 °C than 1907–1960 long-term average (Tielidze et al. 2016b), as certainly affects in negative effects on the stability of glaciers;

6. Also, as some scientist says the some aerial photographs taken on May 17 in the area of origin of the stream, visible clouds like fumaroles (Chernomorets 2014). But in our opinion on this issue it is required to collect additional information.

Finally, in our opinion the cause of the disaster is required much more glacio-geomorphological research.

The Chachi glcier is the third in size after the northern Devdoraki and southern Devdoraki in the Amali river basin. It is a valley-hanging type of glacier morphologically with the northeastern exposition. In recent decades, the rate of glacier retreat is not high and does not reach the average annual index of ∼10 m. It is caused by the fact that the Chachi glacier has a relatively short ice tongue and much larger firn, which is connected to the Devdoraki firn and especially to the firn of the Russia's glacier of Maili. Recent aerial images show, that their firns will soon be split and within a few years they will be formed as independent glaciers.

As for the glacier area, the case is a bit complicated, as it is related to the Georgian-Russian border demarcation line. The fact is that the state boundary line "cuts" the Chachi glacier into two parts and its southwestern part (firn and the small part of the ice tongue) gets within the territory of Russia, and the rest of the northeastern part (main part of the ice tongue)—within the territory of Georgia. Generally speaking, the border is not properly marked and does not coincide with the orographical boundary or the topside separating the firn of the glaciers, it mean, that the borderline is relocated into the territory of Georgia. Because of that the identification of exact area of the glacier is somewhat complicated. Taking into consideration the orographical and morphometrical conditions, we made a contour of the glacier in its natural boundaries the area of which is 1.27 km². The glacier tongue ends at a height of 3350 m above sea level.

The Chkhere River basin is located in the southeastern and eastern slopes of the Kazbegi massif. By the recent data there are four glaciers in this basin with the total area of 7.15 km². By the data of 1960 there were seven glaciers with the total area of 9.86 km². As we can see, with the decrease in number of glaciers their area reduced as well by 27.48%. First of all, it concerns the disappearance of small glaciers.

The glaciers are evenly distributed by morphological types (Fig. 3.83), out of which the biggest is valley glacier of Gergeti.

Due to a small number of glaciers the expositions are of only three directions too, out of which the southeastern glaciers exceed all others by number and area (Fig. 3.84).

The Gergeti (Ortsveri) glacier is the largest in the Chkhere River basin. It flows down from the southeastern slope of Kazbegi massif. Morphologically, the Gergeti glacier is of valley type. Its area is 6.0 km².

The firn basin is located at ∼3900 m above sea level. The glaciers of Gergeti and Devdoraki have a common ice shed; therefore, the area of

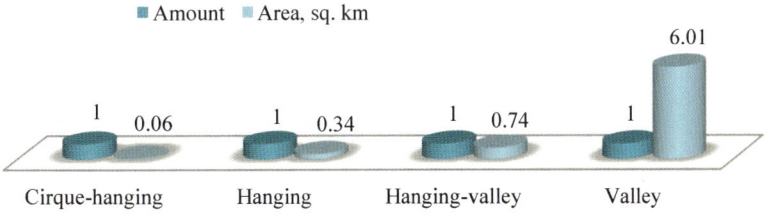

Fig. 3.83 Distribution of the glaciers in the Chkhere River basin according to the morphological types

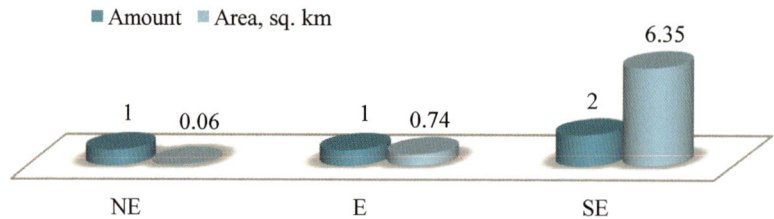

Fig. 3.84 Distribution of the glaciers in the Chkhere River basin according to the exposition

the glaciers, which start in Kazbegi-Mail glacial plateau, is determined approximately. The glacier flows to the south from the firn basin, then turns to the east at Ortsveri peak and creates the icefall. Downward the ice tongue is narrowing gradually and ends at a height 3200 m above sea level. Lateral and terminal stade moraines are well visible in front of the ice tongue in the Ckhere River gorge.

By the data of 1960 the Gergeti glacier area was 6.77 km^2 (Fig. 3.85d). Its tongue was ended with the sharply pointed form at a height of 2880 m above sea level. During the last 54 years the glacier has retreated by ~0.9 km. And according to the individual years, for example, in 2010–2011, we recorded the mechanical breakdown of the glacier tongue, resulting in the glacier retreat by ~40 m. By this index of melting, the Gergeti glacier is in the first place among the other glaciers of the same size in Georgia, if we do not consider the last only 3 years' data of Chalaati glacier, when the glacier's mean annual retreat exceeded the data of Gergeti; however, it should be taken into account, that the Chalaati ice tongue flows down to 1960 m and ends at the elevation below ~920 m compared to the ice tongue of Gergeti. Thus, the Chalaati case may be is not surprising but obviously, in case of Gergeti glacier, we are dealing with the high rates of reduction. We have already considered the reasons that caused this phenomenon, when we discussed the decrease in total area and number of the glaciers in the Tergi River basin.

It is possible to identify the drastic change of the glacier by comparing of the old and modern images and maps with each other (Fig. 3.85).

The Kesia River basin is located on the southern slope of the Khokhi range. By the data

of 1960 there were three glaciers there with the total area of 1.10 km^2. There are two glaciers with the total area of 0.13 km^2 today. The glaciers are of southern and southeastern exposition and morphologically belong to the cirque type of glaciers. These glaciers are located approximately at a height of 3700–4000 m above sea level.

The Mna River basin is located on the southern slope of the Khokhi range and is of meridional direction. By the latest data there are seven glaciers with the total area of 4.82 km^2 in this basin. Due to the high inclination of the slopes all of the seven glaciers are of hanging types morphologically. As for the exposition, the four of the glaciers are of the southern exposition and three of them—of southwestern exposition.

By the data of 1960 there was the same number or seven glaciers with the total area of 9.55 km^2 in the Mna River basin.

The Mna glacier is the largest glacier in this basin. Its area is 2.60 km^2 and the length is 2.50 km. The glacier ends at the height of 3360 m above sea level. Based on the field surveys and the latest aerial images, we can say that the Mna glacier is the first among the same size glaciers in Georgia by its retreating index (we do not imply the large glaciers such as Tviberi and Kvishi, where the retreating index of the same period was even higher).

By the data of 1960 the glacier area was 3.19 km^2 and its length was 4.08 km. Its tongue was ended at the height of 2855 m above sea level. During the last 54 years the glacier area has reduced by 18.49%, which is not a large figure compared to other glaciers. But, if we look through the data on reduction in length of the glacier and variation in ice tongue elevation, we

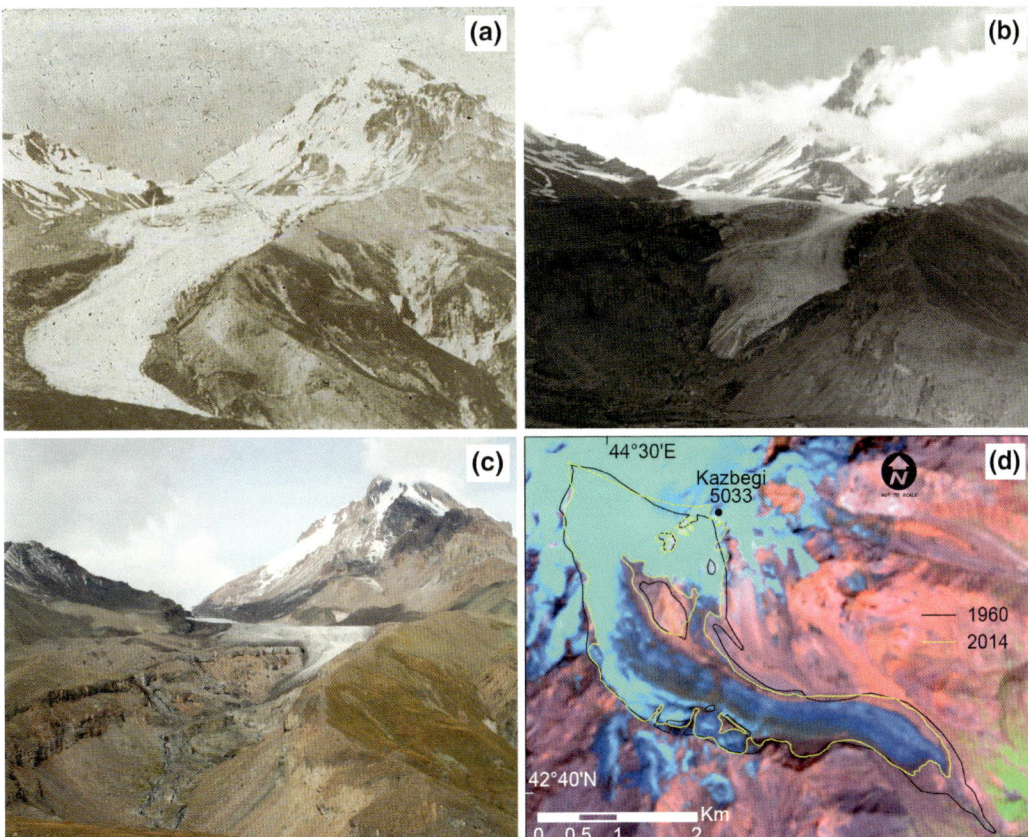

Fig. 3.85 Gergeti glacier reduction, in the years of 1890 (**a**) (*photo by* Museum of Geography at TSU)—1950 (**b**) (*photo by* Museum of Geography at TSU)—2011 (**c**) (*photo by* L. Tielidze). Gergeti glacier reduction, in the years of 1960–2014 (**d**)

can see that here we are dealing with especially high figures. Namely, the glacier tongue is raised by 505 m above sea level and its length is shortened by 1.58 km. The reason is as follows:

The Mna glacier begins in the Kazbegi-Maili plateau. In 1960, after flowing from the firn the glacier created the ∼200 m high icefall, then the ice tongue was much narrowed (∼150–200 m) and was extended along ∼1.4 km in the Mna River gorge. The ice tongue surface was covered with a thin moraine cover, especially its last part. Due to narrow ice tongue the retreating rate is so high. And such a rapid melting away of the ice tongue is a result of a thin moraine cover, which made the ablation process more intensive along with the climatic factors. In addition, as we have noted above, the glacier tongue was too

narrow and was characterized by not a large thickness, and therefore, its existence was short-lived. As can be seen from the aerial images, the main part of the glacier tongue was melted away before 1986, while the rest of the ∼200 m of it was melted away after 1986.

Unnamed glacier, which is the second in size (1.26 km^2) in the Mna River basin, is located between the Mna and eastern Suatisi glaciers. It is fitted so into the watershed located between these two basins, that it is difficult to distinguish, which basin it is more related to. We conventionally related it to the Mna River basin.

The Suatisi River basin is located on the southern slope of the Khokhi range. It is in the first position by the area of modern glaciers in the Tergi River basin. The share of the Suatisi River

Fig. 3.86 Distribution of
the glaciers in the Suatisi
River basin according to
the morphological types

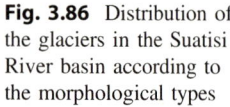
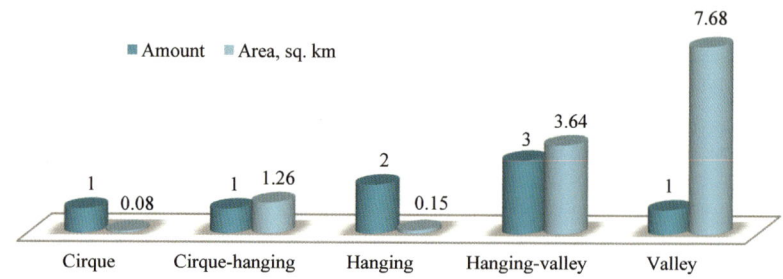

basin glaciers is 36.02% in the total area of the
Tergi River basin glaciers.

Hanging-valley glaciers occupy the first
position by number and the valley glaciers—by
area (Fig. 3.86).

As for the exposition, by the area and number
the first place occupy the glaciers of the overall
southern exposition. Only there are two small
glaciers of eastern and western expositions
(Fig. 3.87).

By the data of 1960 there were nine glaciers in
the Suatisi River basin with the total area of
17.32 km². Morphologically, the five glaciers
with the area of 0.75 km² were of cirque type
and the four of them—of hanging-valley type.

There are eight glaciers with the total area of
12.81 km² by the data of 2014 in the basin.
Glaciers area has the same problem there that we
encountered in the Amali River basin as it was
already mentioned above. Namely, it concerns to
the eastern Suatisi glacier, the firn of which is
"cut" in two by the state border and the northern
part of the firn remains beyond the border. We
made the glacier contours in the natural
boundaries.

Eastern Suatisi glacier is the largest in this
basin. The glacier flows from the southern slope
of the Khokhi range. It has a two-chamber firn.
Morphologically it is a valley glacier, which has
a southwestern exposition (Fig. 3.88c). The glacier area is 7.68 km². In this regard, the eastern
Suatisi glacier occupies the 8th place among the
glaciers of Georgia and by area it exceeds the
glaciers such as Kvishi (7.45 km²), the central
Lekhziri (6.27 km²), Dolra (5.48 km²), et al. It
should be noted that after the Enguri River basin,
in Georgia such a large glacier can be found
neither in the Rioni River basin nor in the Kodori
River basin. Length of the glacier is 4.95 km. Its
ice tongue ends at a height of 2240 m above sea
level. Water runs out of the glacier tongue in two
flows, which is divided by a longitudinal ledge
located in front of the ice tongue. A small lake of
glacial origin with the area of 0.01 km² is located
in the watershed ledge.

By the data of 1960 the glacier area was
11.17 km². During the last 54 years, the reduction of the glacier's area by 31.24% is related to
the reduction of the firn together with the melting
of the ice tongue.

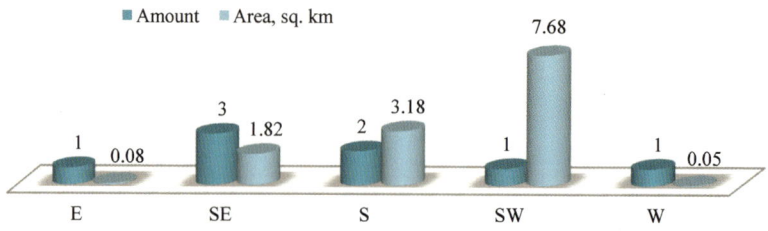

Fig. 3.87 Distribution of the glaciers in the Suatisi River basin according to the exposition

Fig. 3.88 Western (**a**), Central (**b**) and Eastern Suatisi Glaciers. Google Earth Image, 2013

According to the area the next is the Central Suatisi glacier (1.91 km²). It is a hanging-valley glacier with a southern exposition (Fig. 3.88b). After comes the Western Suatisi glacier (1.48 km²), which is the hanging-valley glacier with the southeastern exposition (Fig. 3.88a).

15 glaciers with the total area of 2.39 km² are located in total in the basins of the rivers of **Jimara, Tepistskali, Resistskali, Tsotsoltastskali, Desistskali, and in the headwaters of the Tergi River**. And, there are no modern glaciers at all in the Jimarastskali River basin. By the exposition the 8 glaciers are of overall northern exposition, others are of southern, southeastern, and western exposition. According to morphological types the glaciers of cirque and cirque-hanging types are mainly spread there.

By the data of 1960 there were eight glaciers in the Jimarastskali River basin, 10 glaciers—in the Tepistskali River basin, six glaciers—in the Resistskali River, two glaciers—in the Tsotsoltastskali River basin, six glaciers—in the Desistskali River basin and 11 glaciers in the

Tergi River headwaters. In total, their number was 43. The number of glaciers has increased only in the Tsotsoltastskali River basin, where there were formed four individual glaciers due to the division of two glaciers. But in all other basins the number of the glaciers has decreased. As for the glaciers in the Jimarastskali basin, they have disappeared completely. In total, the number of the glaciers has decreased by 65.11%. They were mainly small glaciers and accordingly, their existence was short-lived.

Juta River basin (Snostskali). The Juta River is a headwater of the Snostskali River, which in its turn is a right tributary of the Tergi River; to the south it is bordered by the main watershed range, to the north—by the watershed of the Juta, Khde, Shanistskali, and Asa basins.

By the data of K. Podozerskiy there were 13 glaciers there with the total area of 3.83 km². B the data of 1960 there were six glaciers with the total area of 1.81 km². By the data of 2014 there are only three glaciers remained with the total area of 0.22 km². All three of them are of

Fig. 3.89 Distribution of
the glaciers in the Khde
River basin according to
the morphological types

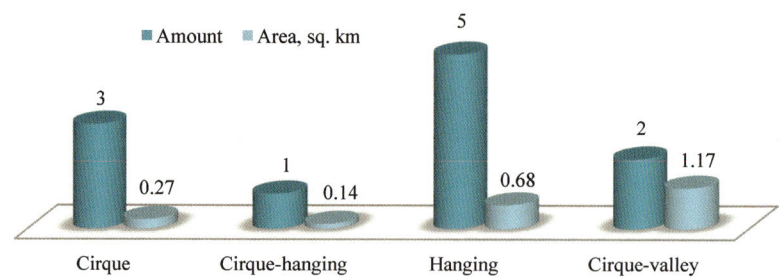

different morphological types—hanging, cirque-hanging, and cirque. Two glaciers are of south-eastern exposition, while the third one—of northeastern exposition. These glaciers are located in the headwaters of the rivers of Veshattskaro and Javartkhokhi Rivers and in the headwater of the unnamed right tributary of the Juta River located between them. Modern glaciers can be no longer found in the Chaukhebi massif.

The **Khde River** is a right tributary of the Tergi River. The gorge is of submeridional direction and maintains the trough form in the headwaters. The Khde River basin is surrounded by the high watershed ranges, where the heights of some of the peaks reach 4000 m (Kuro, Shani, Shino, Bachakhi). According to K. Podozerskiy there were seven glaciers in this basin with the area of 10.85 km^2. By the data of 1960, the number of the glaciers was increased three times, while the area was reduced. By the data of 2014 there are 11 glaciers in the basin with the total area of 2.26 km^2.

Hanging glaciers occupy the first place by the number, and then come the cirque and cirque-valley glaciers. There is only one glacier of cirque-hanging type (Fig. 3.89).

The **Kibesha glacier** is the largest glacier in the Khde River basin. It is a cirque glacier of the northwestern exposition. It is located in the headwater of the gorge on the northwestern slope of the watershed of the basins of the Khde and Juta Rivers. The glacier area is 0.90 km^2, and the length—1.27 km. Its surface is covered with thin weathered materials. In front of the ice tongue the lateral and terminal moraines are remained.

In 1960, the Kibesha glacier area was 2.06 km^2 and its length—1.84 km. In parallel with the reduction in area the glacier was divided into two parts. A small, cirque type glacier of northern exposition was separated from the west side of the main glacier.

The **Bachakhi glacier** is the second in size in the Khde River basin. It is a hanging glacier with the northwestern exposition. It is located on the right side of the gorge, on the western slope of the Shavana ridge. The glacier area is 0.37 km^2.

3.2.10 Glaciers of the Asa River Basin

The Asa River basin is located on the northern slope of the Greater Caucasus. Its name is Arkhotistskali on the Georgia's territory. The Asa River basin is bounded by the Kidegani range to the west, by the Khevsureti range to the east and by the Greater Caucasus watershed range to the southern. It includes the territory of Khevsureti—the historical region, namely, the territory of Arkhoti. Height of some of the peaks in this region (Chamghismaghali, Martinismta, Makhismaghali,   Artsivismaghali)   exceeds 3700 m.

By the data of K. Podozerskiy there were 17 glaciers in the Asa River basin with the total area of 4.14 ± 0.13 km^2. By the topographic maps of 1960, there were nine glaciers in this basin with the total area of 2.59 ± 0.085 km^2. By the data of 2014, there are only three glaciers remained there with a total area of 0.54 ± 0.025 km^2.

As these glaciers are located on the eastern slope of the Kidegani ridge, that is why the two of the glaciers are of eastern exposition and one of the glaciers is of northeastern exposition. Morphologically they are of hanging, cirque, and cirque-valley glaciers.

3.2.11 Glaciers of the Arghuni River Basin

The Arghuni River basin is located on the northern slope of the Greater Caucasus. It includes the Arghuni and Andaki gorges, which have the meridional direction. Modern glaciers are mainly concentrated on the western slope of the Atsunta ridge; only one glacier is located on the northern slope of the Greater Caucasus mountain range, to the east of the place, where the Khevsureti range is separated from the main mountain range.

Although the hypsometric benchmarks of the relief are quite high, the modern glaciation is presented in small scales, and the glaciers are characterized by small sizes.

By the data of K. Podozerskiy there were 10 glaciers in the Arghuni River basin with a total area of 5.43 ± 0.12 km^2. By the data of 1960 topographical maps there were 17 glaciers with the total area of 2.92 ± 0.12 km^2. The glaciers were mainly concentrated on the eastern slope of the Khevsureti range and on the western slope of the Atsunta range. By the data of 2014, there only six glaciers with the total area of 0.43 ± 0.025 km^2. Out of them the glaciers on the eastern slope of the Khevsureti range do not exist anymore.

Melting of the glaciers in this speed is caused not only by the climatic conditions, but by the morphological peculiarities of the relief. The relief of the Arghuni River basin is built by the Jurassic sedimentary rocks, which experience heavy denudation. That is why the Pleistocene glaciation forms are poorly preserved there, where the snow is very well-kept and stored and accordingly, it is one of the important conditions for existence glaciers.

Most of the modern glaciers are of hanging type, only there is one other one—of cirque-hanging type. Most of them are of overall northern exposition.

3.2.12 Glaciers of the Pirikita Alazani River Basin

The Pirikita Alazani River basin is located on the northern slope of the Greater Caucasus and is of latitudinal direction. Glaciers in the region are located on the southern slope of the Pirikita range and (between the Kachu and Diklosmta peaks) and on the eastern slope of the Atsunta range (between Amugho and Tebulo). Here the individual peaks' height is over 3800–4000 m. Only one small cirque glacier is located on the northern slope of the Ruana range.

According to the catalog of K. Podozerskiy there were 23 glaciers with the total area of 19.12 ± 0.32 km^2. By the data of 1960 topographical maps the glaciers were reduced in size; though the number of glaciers was increased up to 36, their area was reduced to 10.48 ± 0.32 km^2. By the data of 2014 there are 20 glaciers in this basin with the total area of 2.42 ± 0.11 km^2.

Cirque glaciers occupy the first position according to the number, followed by the cirque-valley glaciers. There is only one glacier of valley type. The cirque-valley glaciers occupy the leading place by the area. By the smallest area, three hanging glaciers stand out, the total area of which is 0.13 km^2 (Fig. 3.90).

According to the exposition the glaciers of the overall southern exposition are distinguished; they occupy the first position by number. There are seven glaciers of the overall northern exposition. There is only one small hanging glacier of western exposition (Fig. 3.91).

By the data of 1960 the largest glacier was the **Tebulo glacier** in the basin. It was laid up in the deep cirque of the southern slope of the Tebulo glacier. The glacier was of valley type with the

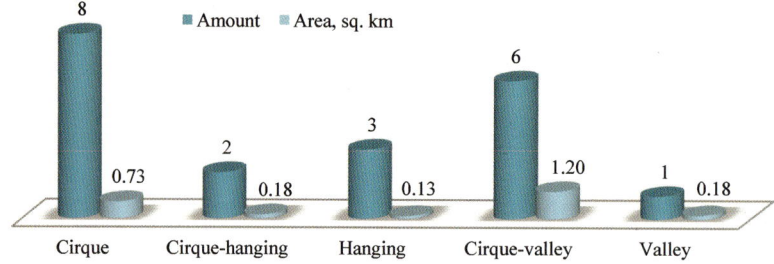

Fig. 3.90 Distribution of the glaciers in the Pirikita Alazani River basin according to the morphological types

Fig. 3.91 Distribution of the glaciers in the Pirikita Alazani River basin according to the exposition

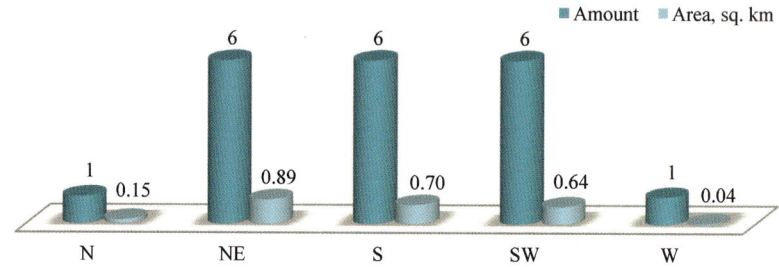

southern exposition. By the data of Kurdghe-laidze (1983), the glacier has retreated by 107 m in the years of 1953–1968. In the relief at the end of the glacial tongue the glacial morphosculptural forms are preserved, which allow to identify the boundaries of the glacier of the LIA maximum. As it can be seen from the aerial images the melting of glaciers was very intensive in the following years as well, resulting in a split of the glacier into two parts.

Now there are two small size glaciers—the cirque and valley glaciers of southern exposition instead of that glacier.

Now in the Pirikita Alazani River basin the largest is the **Dakuekhi glacier**, which is located in the Kvakhidistskali River basin. Its area is 0.30 km², and the length—1.09 km. It is a cirque-valley glacier of the northeastern exposition.

References

Abich H (1865) Isledovanie nastaiashix i drevnix led-nikov Kavkaza (The study of current and ancient glaciers of the Caucasus). Cb, cvedenii o Kavkaze. T. 1, Tiflis (in Russian)

Bhambri R, Bolch T, Chaujar RK, Kulshreshtha SC (2011) Glacier changes in the Garhwal Himalaya, India, from 1968 to 2006 based on remote sensing. J Glaciol 57:543–556

Bolch T, Buchroithner MF, Pieczonka T, Kunert A (2008) Planimetric and volumetric glacier changes in the Khumbu Himalaya 1962–2005 using Corona and ASTER data. J Glaciol 54:562–600

Bolch T, Menounos B, Wheate R (2010) Landsat-based inventory of glaciers in western Canada, 1985–2005. Remote Sens Environ 114(1):127–137. doi:10.1016/j.rse.2009.08.015

Bush NA (1914) O coctaianie lednikov cevernogo sklona Kavkaza v 1907, 1909, 1911 i 1913 godax (About the state of the northern slope of the Caucasus glaciers in 1907, 1909, 1911 and 1913). Izd. RGO. T. 50. vip. 9

Catalog of Glaciers of the USSR (1975) Katalog Ledni-tov USSR. Gidrometeoizdat, Leningrad, vol 8–9, 108 p

Chernomorets SS (2014) New Kazbeg avalanche. Priroda (Nature) 7:67–72, 17 May 2014 (in Russian)

Dinnik NY (1890) Present and old glaciers of Georgia. KOIRGO 14:88–102 (in Russian)

Gobejishvili RG (1995) The evolution of the modern ice age glaciers and mountains of Eurasia in the Late Pleistocene and Holocene. The thesis of doctor of science degree in geography (in Georgian)

Gobejishvili RG (1981) A study of modern relief-forming processes in mountain areas by stereophotogrammetric methods (on the example of Racha in western Georgia). Publishing House of "Metsniereba", Tbilisi (in Russian)

Gobejishvili RG (1989) Glaciers of Georgia. Publishing House of "Metsniereba", Tbilisi (in Russian)

Gobejishvili RG, Tielidze L (2012) The map of the modern and Late Pleistocene (Wurmian) glaciations of Georgia. National Atlas of Georgia. Publishing House of "Cartography", Tbilisi, p 91 (in Georgian)

Gobejishvili RG, Tielidze L, Gadrani L, Latsabidze G, Kumladze R (2013) Glacial-morphological study of the glaciers of the Tergi River basin and evolution of glaciation in Pleistocene. Collected Papers of the TSU Vakhushti Bagrationi Institute of Geography, New Series no 5(84). TSU Publishing House, Tbilisi (in Georgian)

Granshaw FD, Fountain AG (2006) Glacier change (1958–1998) in the North Cascades National Park Complex, Washington, USA. J Glaciol 52(177):251–256. doi:10.3189/172756506781828782

Guo W, Liu S, Xu J, Wu L, Shangguan D, Yao Xi, Wei J, Bao W, Yu P, Liu Q, Jiang Z (2015) The second Chinese glacier inventory: data, methods and results. J Glaciol 61:357–372. doi:10.3189/2015JoG14J209

Inashvili ShV (1975) Ledniki iujnovo sklona centralnogo Kavkaza (The glaciers of the southern slope of the Central Caucasus). Avtoreferat kand Disertacii, Tbilisi (in Russian)

Ivankov PA (1959) Oledinenie bolshogo Kavkazai ego dinamika za godi 1890–1946 (The glaciation of the Greater Caucasus and its dynamics during the years 1890–1946). Izv. VGO. T. 91, Vip. 1 (in Russian)

Kaab A (2002) Monitoring high-mountain terrain deformation from repeated air and spaceborne optical data: examples using digital aerial imagery and ASTER data. ISPRS J Photogramm 57:39–52

Khazaradze RD (1971) Relief, kontinentalnie otlojenia i pleictocenovie oledenenie bac. R, Inguri (Relief, continental sediments and Pleistocene glaciation Enguri River basin). Avtoreferat kand. Disertacii. Tbilisi (in Russian)

Khromova T, Nosenko G, Kutuzov S, Muraviev A, Chernova L (2014) Glacier area changes in Northern Eurasia. Environ Res Lett 9:015003. doi:10.1088/1748-9326/9/1/015003

Kotlyakov VM, Dyakova AM, Koryakin VS, Kravtsova VI, Osipova GB, Varnakova GM, Vinogradov VN, Vinogradov ON, Zverkova NM (2010) Glaciers of the former Soviet Union, in: Satellite image atlas of glaciers of the world—Glaciers of Asia. in: Williams Jr RS, Ferrigno JG (eds) US Geological Survey Professional Paper 1386-F-1, US Geological Survey, Washington, USA, pp 4–5

Kotlyakov VM, Rototaeva OV, Nosenko GA, Desinov LV, Osokin NI, Chernov RA (2014) Karmadon catastrophe: what happened and what we should wait for in future. Moscow: «Kodeks» Publishing House, 184 p (in Russian)

Kovalev PV (1961) Sovremennie i drevnie oledenenie bacceina r. Inguri (Modern and ancient glaciation Enguri River basin). Mat. Kavkaz. Eksped. T. 2. Xarkov. Izd XGU (in Russian)

Kurdghelaidze G (1983) Tusheti, monogr. Publishing House of "Metsniereba". Tbilisi (in Georgian)

Maruashvili LI (1937) O nekatorix faktorax izmenenia lednikogo perioda kavkaza (Some factors of the change the Caucasus glacier period). Izd. GGO. T. 69. Vip. 2 (in Russian)

Paul FR, Barry RG, Cogley JG, Frey H, Haeberli W, Ohmura A, Ommanney CSL, Raup B, Rivera A, Zemp M (2009) Recommendations for the compilation of glacier inventory data from digital sources. Ann Glaciol 50:119–126

Pfeffer WT, Arendt AA, Bliss A, Bolch T, Cogley JG, Gardner AS, Hagen J, Hock R, Kaser G, Kienholz C, Miles ES, Moholdt G, Mölg N, Paul F, Radic V, Rastner P, Raup BH, Rich J, Sharp MJ, The Randolph Consortium (2014) The Randolph Glacier Inventory: a globally complete inventory of glaciers. J Glaciol 60:537–552. doi:10.3189/2014JoG13J176

Podozerskiy KI (1911) Ledniki Kavkazskogo Khrebta (Glaciers of the Caucasus Range). Zapiski Kavkazskogo Otdela Russkogo Geograficheskogo Obshchestva 29(1):200 (in Russian)

Radde GI (1873) Predvoritelnii otchet o puteshestvii d-ra Radde po Kavkazu (Preliminary report on the journey of Dr. Radde the Caucasus). Zap KORGO, kn. 8 (in Russian)

Rashevskiy NN (1904) Cherez Gebivcek (Through pass Gebi), ERGO. T. 3 (in Russian)

Raup B, Kaab A, Kargel J, Bishop MP, Hamilton GS, Lee E, Rau F, Paul F, Soltesz D, Singh Kalsa SJ, Beedle M, Helm C (2007) Remote sensing and GIS technology in the global land ice measurements from space (GLIMS) project. Comput Geosci 33:104–125

Rutkovskaya VA (1936) Sections: Upper Svaneti Glaciers, pp 404–448. In: Transactions of the glacial expeditions, vol 5. Caucasus, the glacier regions. USSR Committee of the II international polar at the centre administration of the hydro-meteorological service. Leningrad

Shahgedanova M, Nosenko G, Kutuzov S, Rototaeva O, Khromova T (2014) Deglaciation of the Caucasus Mountains, Russia/Georgia, in the 21st century observed with ASTER satellite imagery and aerial photography. Cryosphere 8:2367–2379. doi:10.5194/tc-8-2367-2014

Shengelia RG (1975) Opredilenie lednikogo pitania v baseinax rek kavkaza (Determination of glaciers of the Greater Caucasus river basins). Ocherki po fizicheskoe geografiis Kavkaza. Metsniereba, Tbilisi (in Russian)

Stokes CR, Popovnin VV, Aleynikov A, Shahgedanova M (2007) Recent glacier retreat in the Caucasus Mountains, Russia, and associated changes in supraglacial debris cover and supra/proglacial lake development. Ann Glaciol 46:196–203

Tabidze DD (1965) Gomorfologia baseina reka Kodori (Kodori river basin geomorphology), Avtoreferat kand. Disertacii, Tbilisi (in Russian)

The National Environmental Agency (NEA) (2014) General condition of mudflow phenomena in the Tergi River upstream and the report on the assessment of geodynamical conditions of the catastrophic glacial debris flow developed in the Dariali gorge on 17th of May of 2014 (in Georgian). http://nea.gov.ge/

Tielidze LG (2016a) Glacier change over the last century, Caucasus Mountains, Georgia, observed from old topographical maps, Landsat and ASTER satellite imagery. Cryosphere 10:713–725. doi:10.5194/tc-10-713-2016

Tielidze L (2016b) Glaciers catalog of Georgia. Publishing House "Samshoblo", p 116 Tbilisi

Tielidze LG, Kumladze R, Asanidze L (2015a) Glaciers reduction and climate change impact over the last one century in the Mulkhura River Basin, Caucasus Mountains, Georgia. Int J Geosci 6:465–472. doi:10.4236/ijg.2015.65036

Tielidze LG, Lominadze G, Lomidze N (2015b) Glaciers Fluctuation over the last half century in the headwaters of the Enguri River, Caucasus Mountains, Georgia. Int J Geosci 6:393–401. doi:10.4236/ijg.2015.64031

Tielidze LG., Gadrani L, Tsitsagi M, Chikhradze N (2015c) Glaciers dynamics over the last one century in the Kodori River Basin, Caucasus Mountains, Georgia, Abkhazeti. Am J Environ Prot 4(3–1):22–28 (Special Issue: Applied Ecology: Problems, Innovations). doi:10.11648/j.ajep.s.2015040301.14

Tielidze LG, Gadrani L, Kumladze R (2015d) A one century record of changes at Nenskra and Nakra River Basins Glaciers, Causasus Mountains, Georgia. Nat Sci 7:151–157. doi:10.4236/ns.2015.73017

Tielidze LG, Chikhradze N, Svanadze D (2015e) Glaciers amount and extent change in the Dolra River Basin in 1911–1960–2014 Years, Caucasus Mountains, Georgia, observed with old topographical maps and landsat satellite imagery. Am J Clim Change 4:217–225. doi:10.4236/ajcc.2015.43017

Tielidze LG, Lomidze N, Asanidze L (2015f) Glaciers retreat and climate change effect during the last one century in the Mestiachala River Basin, Caucasus Mountains, Georgia. Earth Sci 4(2):72–79. doi:10.11648/j.earth.20150402.12

Tsereteli DV (1959) Izmenenie lednikov iuzhnovo sklona Kavakiona za poslednie 25 let (Changing the southern slope of the Caucasus glaciers over the past 25 years). Coobsh. An. GSSR. T. 21. #6. Tbilisi (in Russian)

Tsereteli DV (1966) Pleistocene deposits of Georgia. Publishing House of "Metsniereba", Tbilisi, 582 p (in Russian)

Vardaniants LA (1930) O novom sposobe podcheta depresii snegovoi granici v sviazi izucheniem stadii otstupania lednikov gornoi gruppi adai-xox v centralnom kavkaze (On a new method estimate depression snow line in relation the study stage glacier retreat mining group Adai-hoch in the Central Caucasus). Izd. RGO. T. 62. Vip. 2 (in Russian)

Abstract

This chapter discusses the variability of the morphological types and expositions of the glaciers of Georgia during the years of 1960–2014. Location of snow and firn lines in 1960 is shown according to the individual river basins and individual glaciers as well.

Keywords

Snow line · Firn line · Glacier aspect

4.1 Morphological Types of the Glaciers

Morphological types of glaciers were formed mainly in the beginning of the nineteenth century, when finished Little Ice Age (LIA) maximum (Gobejishvili 1995). Morphometrical and morpograpical conditions are of special importance in forming the morphological types of glaciers. In order to get certain morphological type of the glacier, a peculiar form of the relief is needed. The Greater Caucasus, by its highly elevated, deeply dissected crest part, varieties of rock lithology and preserved glacial forms due to the old glaciation, conditioned the creation of different morphological types of glaciers. It should also be noted that the formation of morphological type was caused by the glacio-dynamical processes as well. Despite the fact that over the last ~ 200 years, the glaciers of the Greater Caucasus suffer from

degradation, neither new morphological type has been created nor old one has been disappeared; however, this process has great impact on the compound-valley glaciers. For example, in 1960 in Georgia there were 10 compound-valley glaciers with a total area of 155.70 km^2 (Fig. 4.1). Today there are only the seven glaciers of the same type with a total area of 55.04 km^2.

According to the former Soviet Union classification, for today there can be found the glaciers of following morphological types in Georgia: the compound-valley, the simple-valley (with the firn of one or multichamber), the cirque-valley, the cirque-hanging, the hanging, and the cirque types (Fig. 4.2).

According to the number the cirque glaciers occupy the leading place. There are 289 glaciers of the mentioned type with a total area of 40.92 km^2. The second and the third places, respectively, occupy the hanging and cirque-valley glaciers.

L. Tielidze, *Glaciers of Georgia*, Geography of the Physical Environment,
DOI 10.1007/978-3-319-50571-8_4

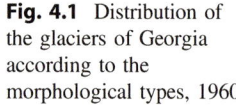

Fig. 4.1 Distribution of the glaciers of Georgia according to the morphological types, 1960

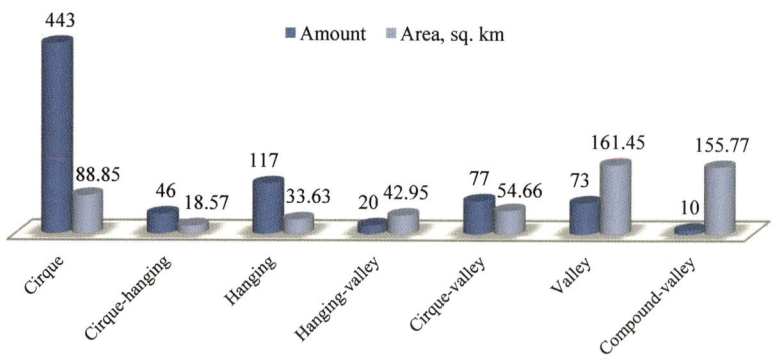

Fig. 4.2 Distribution of the glaciers of Georgia according to the morphological types, 2014

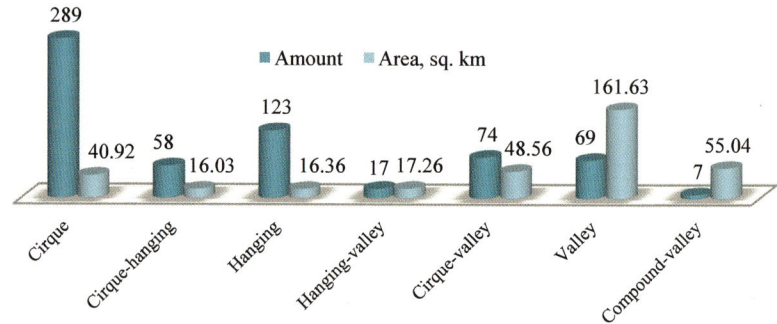

According to number the glaciers of compound-valley type are very few.

According to the area the first place occupy the valley glaciers, the share of which is 45.42% in the total area of the Georgia's glaciers. The second and the third places occupy the glaciers of compound-valley and cirque types, respectively. The smallest area is occupied by cirque-hanging glaciers.

We consider interesting to compare the state of the number and the area of the morphological types of glaciers of 1960 with their current state (Figs. 4.1 and 4.2).

The figures show that we deal with the absolute decrease in the number and area of the most of the morphological types of glaciers. The cirque-hanging, hanging, and valley glaciers do not obey the order of the mentioned changes. In this case, their area was reduced but their number was increased slightly. This can be explained by the fact that as a result of the degradation of the compound-valley and valley glaciers the small, independent, different morphological types of glaciers were formed in some cases.

In Georgia, the simple-valley type of glaciers do not have tributaries, they are started from the firn of one or multichamber. The types of one chamber firns are as follows: Zopkhito, Edena, Dolra, Nageba, Leadashti, Baki, Guli, Marukhi, etc. Glaciers with the multichamber firn have an extensive feeding areal. They are started from the different slopes and morphologically the impression is that as if it is a single firn. In fact, it is composed of several cirque forms. Such glaciers are as follows: Adishi, Boko, Shdavleri, Eastern Suatisi, etc.

Cirque types of glaciers are distributed in every river basins of Georgia. In some basins there are only these types of glaciers (Kelasuri, Khobistskali, Amtkeli, Magana, and Khaishura).

Hanging types of glaciers are related to the highly elevated massifs with steep slopes. By percentage this type of glaciers are spread in the Tergi River basin in highest amount.

Ice tongues of the morphologically different types of glaciers are ended at the different elevations. Hypsometrically lowest (1960 m above sea level) is a compound-valley glacier of

Chalaati. Cirque and hanging glaciers are at the highest elevations.

There were very few compound-valley glaciers in the second half of the 19th and the first half of the 20th centuries, but according to the area they occupied the first place. The following largest glaciers belonged to this morphological type: Tviberi, Tsaneri, Lekhziri, Chalaati, Ushba, Laila, etc. All of the glaciers were formed by joining two or more glacial flows. Their degradation resulted in the fact that the simple-valley glaciers occupy the first place according to the area.

4.2 Exposition of the Glaciers

The center of the glaciation is a Greater Caucasus mountain range in Georgia; its southern slope is located in the territory of the western Georgia and it is represented by its both southern and northern slopes in the eastern Georgia. Naturally, the glaciers of all expositions can be found in Georgia (Fig. 4.3).

By the number the glaciers of the north-western exposition are in the first place, followed by glaciers of southern and northern expositions. The glaciers of eastern exposition are very few in number.

According to the area, the glaciers of southern exposition are in the first place. Their share is 20.47% in the total area of the glaciers of Georgia, followed by the glaciers of southwestern exposition—20.34%. The glaciers of eastern exposition are in the last place—6.75%.

The glaciers of overall northern exposition (N, NE, and NW) are in the first place by the number. Their share is 41.13% in the total number of the glaciers of Georgia and the share of the glaciers of overall southern exposition is 37.67%.

The glaciers of southern exposition are in the first place according to the area and their share in the total area of the glaciers of Georgia is 54.99%. As for the glaciers of northern exposition, they occupy 31.42% of the area. Glaciers of eastern and western expositions are equally distributed (Fig. 4.3).

Such a large amount of the glaciers of the northern exposition is conditioned by caused by the meridional and sublatitudinal direction of the branch-ranges of the Greater Caucasus. Due to their morphometrical and morpographical conditions, existence of small glaciers (Lechkhumi, Chkhalta, Bzipi, Kodori, Shdavleri, and other ranges) are available on the northern slopes of only these mountain ranges.

By the data of 1960, the expositions of the glaciers by number and area are as follows: the share of the glaciers (292 glaciers) of the southern exposition is 37.20% in the total number and 50.10% in the total area (278.75 km^2) of the glaciers of Georgia. There were 348 glaciers of the overall northern exposition (45.3% of the whole amount) with the total area of 161.85 km^2 (29.1%) (Fig. 4.4).

Analysis of given materials (Figs. 4.3 and 4.4) shows that during the last 54 years the number of glaciers and their distribution by area in line with the time has not changed much in percentage and there were in them.

4.3 Location of Snow and Firn Lines

Study of snow and firn lines is of interest of many fields of geographical science. Their location gives a clear idea of the nature of the glaciations of any region. Changes in the location of these lines affect the rivers' glacial runoff and

Fig. 4.3 Distribution of the glaciers of Georgia by exposition, 2014

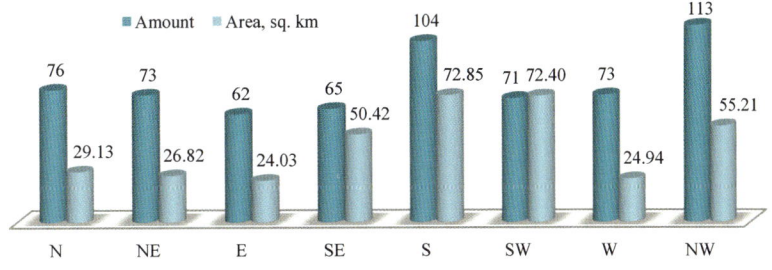

Fig. 4.4 Distribution of the glaciers of Georgia by exposition, 1960

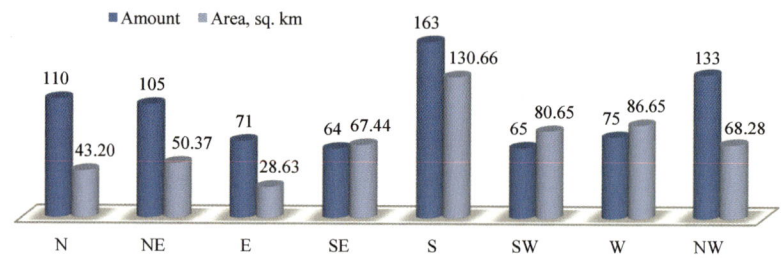

the character of the nival-glacial process, and in its turn, their location is depended on the climate change. The snow line is the boundary of the earth's natural zones in vertical and horizontal extent. Therefore, this research is of theoretical and practical importance.

Altitudinal location of snow and firn lines is very important in determining the nature of the modern glaciations and regime of the glaciers in Georgia. There are different points of view in Geographical literature on definition of snow and firn lines (Kalesnik 1963; Tronov 1972; Shumskiy 1964 etc.). Having analyzed all opinions Kotlyakov (1968) gave a complete definition of snow and firn lines, which we have used in our work.

The location of the modern firn line of the greater Caucasus and individual river basins has been identified mainly by the aerial images, as well as topographic maps of the 1960s and partly with the help of literary sources. We identified the firn line location by *Gefer's method*. Choosing this method was conditioned by two

reasons: (1) the data obtained on location of the firn line do not differ from the data obtained by other methods and (2) while calculating firn line depression, we rely on the forms created by glaciers, stade, and microstade moraines. In this case, this method is more convenient.

Location of firn line for the river basins and their tributary river basins in Georgia was early determined by Reinhardt (1916) according to the one-mile topographic maps. By using the topographic maps of different periods R. Gobejishvili calculated the firn line location for the years of 1946–1950 and 1960–1965 (Tables 4.1 and 4.2).

Table analysis shows that the height of the firn line increases from west to east. The highest (∼3500 m) firn line is located in the river basins of the eastern Georgia.

The firn line height within the individual river basins is not similar. In some of the river basins in the Central Caucasus (Dolra, Mulkhura, and Enguri headwaters) are located higher than in the basins located in its east (Rioni River basin). To

Table 4.1 The height of the firn line in the river basins of Georgia in 1946–1950

No.	Name of the basin	Firn line location (∼m)
1	Bzipi	3030
2	Kodori	3070
3	Enguri	3320
4	Rioni	3380
5	Liakhvi	3480
6	Aragvi	3500
7	Tergi	3450
8	Asa	3460
9	Arghuni	3460
10	Pirikita Alazani	3480

Table 4.2 Variability of firn line elevation ($\sim$ m, above sea level) in 1890–1960

Name of the basin	A. Reinhardt	R. Gobejishvili		Firn line depression	
	By the maps of 1890–1900	By the maps of 1946–1950	By the maps of 1960	By the maps of 1890–1960	By the maps of 1946–1960
Chkhalta	2890	2990	3030	140	30
Klichi	2860	3040	3070	210	30
Ghvandra	2770	2990	3040	270	40
Sakeni	3090	3090	3150	60	60
Mean in the Kodori basin	2900	3030	3070	170	40
Nenskra	3080	3120	3170	90	50
Nakra	3190	3150	3230	40	80
Dolra	3210	3360	3380	170	20
Mulkhura	3200	3380	3400	200	20
Enguri headwaters	3200	3410	3450	250	40
Mean in the Enguri basin	3175	3285	3325	150	40
Edenura	3180	3350	3370	190	10
Zopkhitura	3180	3260	3330	150	70
Chveshura	3140	3315	3370	230	55
Notsarula	3140	3305	3350	210	45
Chanchakhi	3370	3460	3490	120	30
Mean in the Rioni basin	3200	3340	3380	180	40
South slope of the Greater Caucasus	3090	3220	3260	170	40

our opinion, the reason is the high hypsometrical location of the Central Caucasus and the high inclination of its southern slope.

The firn line height in the glacier and in the basin itself is located at different heights within one river basin. The firn line height is always lower on the surface of the large glaciers than in the river basins. For example, in the Chveshura River basin the firn line is located at a height of $\sim$ 3370 m, while the firn line height is $\sim$ 3150 m in the Kirtisho—the largest glacier in the basin. In the Mulkhura River basin the firn line is located at about $\sim$ 3380 m, while in the Lekhziri and Tviberi glaciers the firn line heights are relatively $\sim$ 3120 and $\sim$ 3150 m.

Morphological and morphometrical characteristics of the relief and the morphological type

of the glacier are of great importance for the location of the firn line. The firn line is located lower in the large glacier than in the small glacier. The highest location of the firn line is characteristic for hanging type of glacier (the Khokhi and Atsunta ranges).

It is known that along with the other factors the exposition of the slopes is of great importance in the formation of glaciers. In such ranges, where the glaciers are represented on the northern and southern slopes, the firn line is always located higher on the southern slope than on the northern slope. For example, the firn line is located at a height of $\sim$ 3230 m on the northern slope of the Svaneti range, while at a height of $\sim$ 3350 m—on the southern one. The firn line is located at a height of $\sim$ 3650 m on the southern slope of the Khokhi range and at a height of ∼3580 m—on the northern one.

In the Kodori and Samegrelo ranges the firn lines are located lower, at a height of ∼3050–3100 m. It is due to the abundance of atmospheric precipitation in winter and orographic conditions of the relief.

In 1890–1960, the firn line location underwent the changes (Table 4.2). In the above table, only the southern slope of the Greater Caucasus within the western Georgia is considered. On the southern slope of the Greater Caucasus the firn line was raised by 160 m in average from 1890 to 1960, or by 2.3 m per year, and by 2.5 m per year—in the years of 1946–1960.

Since the firn line depression is same in the large river basins, we can conclude that in this period there was an equal variation in physical–geographical conditions in Georgia.

In our opinion, direct observation materials on the snow line location in the Greater Caucasus does not exist in the literature, but it can be determined by means of indirect methods. Kotliakov (1968) points out: "It is an imprint of the lower level of the hyonosphere on the surface of the earth's relief in the real conditions."

We believe that the lower boundary of the hyonosphere is a mean height of the zero isotherms of the July–August months. Zero isotherms can be determined in two ways: (1) By means of the temperature vertical gradient and (2) According to the materials of aerological observations.

The work on the "Climate and climatic resources of Georgia" (Gavasheli 1971) shows the height of the zero isotherm in the Central Caucasus by help of the air temperature vertical gradient. According to Gavasheli (1971) in July–August, the height of the zero isotherm is ∼4100–4200 m. Similar data have been obtained by Kordzakhia (1967) and Inashvili (1975) for the Enguri River basin.

Based on the materials of aerological observations, the air temperature distribution in the free atmosphere was studied by R. Gobejishvili for many stations in the former Soviet Union. In the Greater Caucasus the average height of the air zero isotherm in July–August is equal to ∼4120

m over the mineral waters (Northern Caucasus), ∼4465 m—over Sokhumi and ∼4500 m over Tbilisi (Kvaratskhelia 1964). We determined the height of the zero isotherm by average data of these data for the Central Caucasus, which is equal to ∼4360 m.

Above data shows that in the Central Caucasus in the Enguri River basin the height of the zero isotherm ranges within ∼4200–4480 m. As we can see the data received according to the air temperature vertical gradient is lower than the height of the zero isotherms in the free atmosphere, which is caused by the influence of the relief of the Greater Caucasus.

Shumskiy (1964) points out, that the snow boundary it is the lower border of the snow-feeding district or zone (namely, the "permanent snow zone", recrystallization).

Issues of ice formation zones for the glaciers of the Greater Caucasus are poorly discussed in the literature. During the second International Geophysical Period the complex studies were conducted in the Elbrus peak, by the results of which the ice formation zones were identified; regelation-recrystallization firn's lower boundary was located at the height of 4200 m an average (Gobejishvili 1995).

In September 1987, In Svaneti region the complex expedition of Georgian and Russian glaciologists was held, the program of which included the study of the firn plateau located above the Adishi glacier. Observations showed that the regelation-recrystallization firn was started at a height of ∼4200–4300 m (Gobejishvili 1989).

Tushinskiy (1968) having used the meteorological stations located at different heights, determined the height of the "Level 365" for the territory of the former Soviet Union. For Central Caucasus the "Level 365" is located at the height of ∼4200 m in average and for the whole Caucasus—at the height of ∼4300 m.

Analysis of the material above allows us to determine the snow line height in the Central Caucasus. It ranges within the ∼4200–4400 m in July–August. Taking into account the sublatitudinal direction of the Caucasus, which is stretched from the northwest to the southeastern direction and is located among the 42° to 44° of the North Latitude, and then the snow line height in the Western, Central and Eastern Caucasus will be at different heights. For example, the Enguri River basin (the Central Caucasus) is located in the 43° of the North Latitude and the snow line is at the height of ~ 4200–4400 m.

Western Caucasus is located northward from the Central Caucasus, and the Eastern Caucasus—southward. Therefore, the height of the snow line will be lower in the Western Caucasus than in the Central Caucasus, and in the Eastern Caucasus it will be even higher. It is known that the air temperature changes by ~ 0.5° per 1° of latitude from the equator to the north, and the temperature falls by ~ 0.5°–0.7° per 100 m height vertically. Hence, based on the aerological data [Tbilisi (Southern Caucasus), Mineral Waters (Northern Caucasus)] can be concluded that the snow line is located at the height of ~ 4100–4200 m in the Western Caucasus, and at the height of ~ 4400–4500 m in the Eastern Caucasus.

Average multiannual altitudinal location of the zero isotherms in the free atmosphere for the north hemisphere—in the latitude of the city of Tbilisi is provided by Davitaia and Tavartkiladze (1982). According to their data, in July–August the height of the zero isotherm changes according to the geographical conditions of the underlying surface and ranges within the ~ 4150–4300 m.

Based on the researches we have conducted a comparative analysis of snow and firn lines, which clearly shows that the difference between snow and firn lines in the Greater Caucasus is irregular. This difference is ~ 1000–1200 m within the western Georgia, while it is ~ 800–1000 m in the eastern Georgia.

On the basis of the conducted study, we can conclude that in 1960 the average height of the firn line in the Western Caucasus was ~ 3050 m, in the Central (within the western Georgia)—~ 3350 m, and in the Eastern Caucasus—~ 3450 m. In 1890–1960 in the southern slope of the Greater Caucasus the firn line raised by 170 m

and within the 1946–1960 period—by ~ 40 m. The snow line is defined in different ways; it was located at the height of ~ 4100–4200 m in the Western Caucasus, at ~ 4200–4400 m—in the Central, and at ~ 4400–4500 m—in the Eastern Caucasus. The firn and snow lines locations rise from west to east, which is caused by the sub-latitudinal direction of the Caucasus together with the other factors.

References

Davitaia FF, Tavartkiladze KA (1982) Problemi barbi s gradobitiem (Problems with hailstorms). Publishing House "Metsniereba", Tbilisi (in Russian)

Gavasheli M (1971) Klimat i klimaticheskie resursi gruzii (Climate and climatic resources of Georgia). Gidrometeoizdat (in Russian)

Gobejishvili RG (1989) Glaciers of Georgia. Publishing House "Metsniereba", Tbilisi (in Russian)

Gobejishvili RG (1995) The evolution of the modern ice age glaciers and mountains of Eurasia in the Late Pleistocene and Holocene. The thesis of doctor of science degree in geography (in Georgian)

Inashvili ShV (1975) Ledniki iujnovo sklona centralnogo Kavkaza (The glaciers of the southern slope of the Central Caucasus). Avtoreferat kand. Disertacii, Tbilisi (in Russian)

Kalesnik SV (1963) Ocherki glatsiologiy (Features of glaciology), geografgiz, Moscow (in Russian)

Kordzakhia R (1967) Enguri and Tskhenistskhali River basins climate features within Svaneti. Acts Georgian Geogr Soc IX–X:110–125 (in Georgian)

Kotlyakov VM (1968) Snejnii pokrov zemli i ledniki (Snow cover and glaciers). Gidromedizdat (in Russian)

Kvaratskhelia IF (1964) Aerologicheskie isledovanie zakavkazie (Aerological research the Transcaucasus). Gidrometeoizdat (in Russian)

Reinhardt AL (1916) Snejnaya granica Kavkaze (The snow line in the Caucasus), Izvestia Kavkazskogo otdela Imperatorskogo Russkogo Geograficheskogo Obshchestva, T. 24. vol 3 (in Russian)

Shumskiy PA, Krenke AN, Zotikov IA (1964) Ice and its changes. In: Solid earth and interface phenomena, vol 2, Research in geophysics. MIT Press, Cambridge, pp 425–460

Tronov MV (1972) Faktory oledeneniia i razvitiia lednikov. TSU, Tomsk, p 235 (in Russian)

Tushinskiy GK (1968) Glaciation rhythms and tenderness in the Caucasus in historical times. In: Elbrus glaciation (in Russian)

Abstract

This chapter discusses the dynamics of the glaciers of Georgia by periods of individual decades since the LIA to the present day. The rates of old glaciers are reconstructed based on the micro-stade moraines; also the topographic maps of the 19–20th centuries and the modern satellite images (Landsat/ASTER) are used. Valley glaciers are grouped into different categories according to their retreat speeds.

Keywords

Glacier dynamics · Little ice age · Moraine

5.1 Dynamics of the Glaciers in 1890–1960

Maps and aerial images of different years are needed to describe the dynamics of glaciers. In 1880–1890 the large-scale topographic maps of the Caucasus were compiled by using the plane table surveying method. But the basis for the maps drawn up in 1946–1950 and 1960 are the aerial photo materials of 1946 and 1958–1960. The mentioned materials are used by the researchers of the Caucasus, in the works of which the morphometric and morpographic characterization of the glaciers are presented. We used the researches conducted by R. Gobejishvili on the basis of 1960 topographical and aero-photo materials. We identified the number and area of glaciers for individual river basins of Georgia and compared with the data of the previous researchers (Table 5.1).

Table analysis shows that the authors give different data about the area and the number of glaciers both according to the individual river basins, and for the entire territory of Georgia. It is difficult to estimate the dynamics of the glaciation in time and space according to the given materials. Data of P. Ivankov and R. Gobejishvili for the glaciers of the southern slopes of the Greater Caucasus are obtained from the topographic maps of the same period (1946–1950) and naturally the area and the number of the glaciers should be almost equal.

Nevertheless according to the data of P. Ivankov the area of the glaciers is higher by 2% and their number—by 9.7%. Especially significant difference is in the number of glaciers (122 glaciers). In our opinion, the reason for the

© Springer International Publishing AG 2017
L. Tielidze, *Glaciers of Georgia*, Geography of the Physical Environment,
DOI 10.1007/978-3-319-50571-8_5

Table 5.1 Changes of the glaciers of Georgia according to the river basins

Basin name	Podozerskiy (1911) based on 1880–1910 topographic maps		Ivankov (1959)		USSR Catalog (1975), based on 1955–1960 aerial images	
	Number	Area (km^2)	Number	Area (km^2)	Number	Area (km^2)
Bzipi	10	4.03 ± 0.085	45	9.66	16	7.80
Kelasuri					3	1.50
Kodori	118	73.20 ± 1.55	174	87.54	141	60.00
Enguri	174	333.03 ± 4.57	210	349.58	250	288.30
Khobistskali					7	1.60
Rioni	85	78.12 ± 1.61	121	84.92	124	62.90
Liakhvi	12	5.15 ± 0.13	34	10.41	22	6.60
Aragvi	3	2.21 ± 0.04	35	2.64	6	1.60
Tergi	63	89.12 ± 1.22			129	72.13
Asa	17	4.14 ± 0.13			3	1.12
Arghuni	10	5.43 ± 0.12			14	1.70
Pirikita Alazani	23	19.12 ± 0.32			40	8.90
Total	515	613.55 ± 9.80			755	514.15

Basin name	R. Gobejishvili, by the maps of 1946–1950		R. Gobejishvili, by the maps of 1960	
	Number	Area (km^2)	Number	Area (km^2)
Bzipi	18	9.36	18	9.90 ± 0.20
Kelasuri			1	0.26 ± 0.015
Kodori	144	81.05	160	63.73 ± 1.63
Enguri	187	349.99	299	323.70 ± 5.72
Khobistskali	7	2.36	16	1.12 ± 0.07
Rioni	100	81.64	112	76.77 ± 1.66
Liakhvi	17	8.22	16	4.27 ± 0.13
Aragvi	3	1.20	3	0.88 ± 0.03
Tergi	106	75.66	99	67.01 ± 1.33
Asa	10	4.02	9	2.59 ± 0.085
Arghuni	17	7.08	17	2.92 ± 0.12
Pirikita Alazani	25	6.66	36	10.48 ± 0.32
Total	634	627.24	786	563.70 ± 11.31

difference is as follows: P. Ivankov did not carry out the decoding of the cartographic and aerial image materials, therefore, the numerous small size snow covers were considered as glaciers (in the gorges of the rivers of Bzipi, Kodori, Aragvi, and Liakhvi) by him.

The comparison of the data obtained from the maps of 1946–1950 by Podozerskiy (1911) and R. Gobejishvili showed that in this period the area of the glaciers increased by 2.1% and their number—by 22.0%. It is known that during the last ∼200 years the areas of the glaciers decrease in the Caucasus and in the earth's surface in general, therefore these data are contrary to the general regularities of the development of the glaciation. In our opinion the increase of the area of the glaciers is caused by the following reasons: it is known that the data of K. Podozerskiy are obtained from the maps of 1880–1890 and their detailed analyses showed the following: (1) The glaciers located in the branch-ranges of the Caucasus are missed in the work of K. Podozerskiy; (2) The difficult is to access firn valleys of the valley glaciers of the southern slope of the Caucasus are depicted very incompletely in the maps of that period, which, of course caused the reduction of the glacier areas in the works of K. Podozerskiy. Researchers (Panov and Vinogradov) of glaciers in the Caucasus point to this lack, who has tried to specify the data of Podozerskiy (Gobejishvili 1995).

Comparison of the glacier areas in the river basins shows that the glacier areas in the river basins of the western Georgia have increased by 7.0%, while in the river basins of the eastern Georgia—have reduced by 17.6%. The reduction of the glacier areas in the eastern Georgia well reflects the dynamics, which is typical for the Caucasus glaciers. It should be noted that the river gorges of the eastern Georgia are easy to access for the plane table survey. Naturally, the glaciers are more completely and accurately displayed on the map, rather than the firn valleys of the hard-to-access glaciers of the Western and Central Caucasus.

Comparison of the data obtained by R. Gobejishvili according to the maps of the year of 1960 with the materials of the former Soviet Union Catalog of glaciers (1975) shows that by the data of R. Gobejishvili the area of the glaciers increased by 8.1%. This figure does not correspond to reality (Table 5.1). It appears that the increase in the area of the glaciers is observed mainly in the western Georgia's glaciers (11.1%), while the area of eastern Georgia's glaciers reduced by 5.6%. The data on the glaciers of Georgia in the catalog are obtained from the maps of the years of 1946–1950, which are modified according to the aerial images of 1955–1960 years. Therefore, the data of the western Georgia's glaciers in the catalog should be nearer to the R. Gobejishvili's data or should be allocated among the data obtained by the maps of 1946–1950 and 1960.

The analysis of the same table shows that there is a very big difference between the data of R. Gobejishvili and catalog in the Enguri and Rioni River basins. The Enguri River basin's glaciers have been studied by and Khazaradze (1971). According to G. Khazaradze's data there are glaciers with the area of 320.4 km^2, and according to Khazaradze—319.6 km^2. Both researchers' data are obtained from the maps of the years of 1946–1950 and they amended them by the aerial images of the years of 1955–1960 and according to the field work materials. Against the background of the researchers' data, the total area of the glaciers (288.3 km^2) in the Enguri River basin according to the catalog becomes doubtful. According to the researches of Vartanov (1978) the area of the glaciers in the Enguri River basin is about 300.0 km^2.

In our view the reason for such a big difference in the glaciers area is the errors made during the catalog compilation, particularly the configurations of the glaciers are depicted in the maps of the years of 1946–1950 by the authors visually, according to the aerial images of the years of 1955–1960 and after they conducted the cartometric works. The use of such a tool in studies of glaciers resulted in their distortion. By the data of Gobejishvili (by the maps of 1960) the number of the glaciers is more by 4.1% than it is shown in the catalog (1975). This difference is mainly caused by the fact that the catalog does not include the small glaciers located in the branch-ranges.

Comparison of data, by Podozerskiy (1911) and Gobejishvili (by the maps of 1960), shows that the area of the glaciers was decreased by 8.12 ± 1.80% in Georgia over the period of 1911–1960. The area of the glaciers in the Enguri and the Rioni River basins were reduced, respectively, by 2.80 ± 1.57 and 2.50 ± 2.11%, while in the Tergi River basin—by 24.80 ± 1.68%; reduction in the glacier areas in the Tergi gorge well reflects the overall picture, which is typical to the glaciers of Georgia. It should be noted that the ice tongues of the valley glaciers have been reduced in average by 16.0% after the LIA maximum.

The area of the glaciers of the eastern Georgia was reduced by 30.11 ± 1.90%. Such a great reduction is caused by the fact that small glaciers with the area reduced in average by 49.0% were distributed in the basins of the rivers of Aragvi, Pirikita Alazani, Arghuni, Asa, Khde, and Juta. At the same time the area of the glaciers of Kazbegi massif (Khokhi range) was reduced by only 16.8%.

Against the background of joint reduction in the area of the glaciers in the Enguri and Rioni River basins, the area of the glaciers was increased in some of the tributary basins in the years of 1890–1960 (Dolra, Mestiachala Enguri headwaters, Zopkhitura). Increasing in the area of the glaciers is caused by the fact that the inaccessible glaciers located in these basins are depicted incompletely in the old maps (Table 5.2).

The table clearly shows that the area of all of the glaciers by the data of the years of 1946–1960 is greater than the data of 1890. Its reason we have explained above, and practically it is impossible to make correction. The area of the glaciers was reduced by 11.3% from 1946 to 1960 and increased by 5.0% in the years of 1890–1960.

To have clear idea of the dynamics of the glaciers of Georgia, R. Gobejishvili had compared the valley glaciers, the firn basins of which are easy to conduct the plain table survey and therefore, they are well depicted in the maps of 1890 (Table 5.3).

Table analysis shows that the area of the glaciers reduced by 14.2% in the years of 1890–1960. These data clearly reflect the regularities,

Table 5.2 Changes of the valley glaciers in 1890–1946–1960

Glaciers name	Glaciers area ($\sim km^2$)		
	1890	1946	1960
Leadashti	3.90	4.84	4.30
Dolra	6.03	9.40	8.01
Chalaati	11.25	13.82	12.81
Lekhziri	38.49	40.44	35.96[a]
Shkhara	3.26	6.14	5.60
Namkvani	3.64	4.40	3.70
Koruldashi	2.67	4.38	3.23
Zopkhito-Laboda	5.90	6.08	5.90
Boko	4.90	5.18	4.62
Total	80.04	94.68	84.13

[a]The Lekhziri glacier area does not include the small glaciers areas separated from it during this period

Table 5.3 Changes of the valley glaciers in 1890–1946–1960

Glaciers name	Glaciers area ($\sim km^2$)		
	1890	1946	1960
Shdavleri	3.16	2.74	2.48
Khalde	13.37	12.80	10.90
Nageba	7.70	7.06	6.07
Gergeti (Ortsveri)	7.13	6.80	6.77
Central Suatisi	3.35	3.06	2.74
Western Suatisi	3.02	2.80	2.50
Mna	4.10	3.24	3.19
Total	41.83	38.50	34.65

which was typical to the glaciers of Georgia during these 70 years.

Over the 1890–1960 period the number of glaciers of Georgia increased by 51.7%. The number of glaciers increased almost in the all of the river basins (excluding the Asa River basin). Increase in the number of glaciers is caused by their division during the degradation of the glaciers.

5.2 Dynamics of the Glaciers in 1946–1960

Studying the dynamics of the glaciers in the period 1946–1960 is possible by comparing the large-scale topographic maps, which are drawn based on the aerial images made in the same period and have been published later (1950–1965).

Though the cartographic material is very informative, their freely acceptance as a basis and conduct the cartometric analysis without special study of the maps would resulted in a great mistake. Therefore, we carried out the desk decoding of the aerial images of the years of 1958; specified the number of the glaciers in the maps and the configuration as well, especially for the glacier tongue areas. Having convinced of the reliability of the cartographic data, we carried out a cartometric measurements. The obtained data allows us to consider the dynamics of the glaciers according to the separate basins (Table 5.1).

Table analysis shows that in the years of 1946–1960 the area of the glaciers reduced by 11.4% and the number of glaciers increased by 8.5%. Glaciers reduction according to the river basins is as follows: Enguri (8.4%), Rioni (8.0%) and Tergi (11.7%). Glaciers area in the Kodori and Bzipi River basins was reduced by 20–30%. Glaciers area in the river basins of Liakhvi, Aragvi, Asa, and Arghuni was reduced by 50% and more, which was caused by the strong degradation of small glaciers.

The number of glaciers was increased everywhere in the western Georgia, which is caused by their division during retreating. Decrease in the number of glaciers in the river basins of the eastern Georgia is mostly caused by the disappearance of small glaciers, with the exception only of the Pirikita Alazani River basin, where the number of glaciers and their area has increased. The main reason for this increase is that not all of the glaciers are depicted in maps of the years of 1946–1950.

The reduction of the valley type of glaciers went on differently in the years of 1946–1960 (Tables 5.2, 5.3 and 5.4). Only 24 glaciers are considered in the tables the total area of which was decreased by 27.07 km^2 (11.6%). The area of the compound-valley glaciers (Tviberi, Lekhziri, Tsaneri) reduced greatly. Main reason for glaciers reduction was the separation of small glaciers from them. Comparison of the topographic maps of different periods showed that the

Table 5.4 Changes of the valley glaciers in 1946–1960

Glacier name	Glaciers area ($\sim$km^2)		
	1946	1960	Difference
Marukhi	1.96	1.80	−0.16
Klichi	1.20	1.00	−0.20
Tviberi	30.72	24.72	−6.00
Kvitlodi	14.36	12.08	−2.28
Tsaneri	32.14	28.30	−3.84
Adishi	10.40	10.50	+0.10
Laila	6.10	5.96	−0.14
Buba	3.90	3.75	−0.15
Total	100.78	88.11	−12.67

Table 5.5 Changes of the glaciers in 1946–1960

Basin name	Glaciers stationary condition ±1.0 m/year	Glaciers retreat (m/year)				Advance of the glaciers (m/year)			Total
		1–10	11–20	21–40	More than 40	1–10	11–20	21–40	
Bzipi	4	4	2	2	–	2	1	1	16
Kodori	19	32	34	8	9	9	2	2	115
Enguri	18	24	26	30	23	5	7	5	138
Rioni	24	16	17	9	5	3	9	–	83
Liakhvi	3	–	2	2	1	–	–	–	8
Tergi	15	18	5	3	1	1	2	–	45
Total	83	94	86	54	39	20	21	8	405

reduction in the area (4.48 km^2) of the Lekhziri glacier was caused by the separation of the Dalakora glacier from it. The four glaciers were separated from the main body of the Tviberi glacier, two glaciers—from the Kvitlodi glacier, and three glaciers—from the Tsaneri glacier. The Lailchala and Zopkhito-Laboda glaciers were divided.

Changes of the glaciers in length. As we saw, the area of the glaciers of Georgia was decreased in the years of 1946–1960. Mainly the glaciers' tongues were retreating. Nevertheless, against the background of common retreat some of the glaciers maintained the stationary state, while some of them even advanced.

In order to investigate the glaciers variability R. Gobejishvili developed a new way to use the cartographical method as follows: the ends of the glacier tongues were measured from the fixed grid point in the topographical maps of the different periods with the same scale and similar surveying basis. During the measurements the morphological features of the relief, the glaciers direction in the valley, configuration of its tongue, etc., were taken into account. All these signs were specified by decoding of the aerospace images.

By this method, 405 glaciers have been studied in the Greater Caucasus. The obtained numerical data have been put in sequence according to the similar signs. These data were divided into groups, taking into account the speed of the glaciers. Eight groups were used at

this stage. Out of them the glaciers was combined in four groups that have been advanced, and a group was separated, which included the glaciers, the movement speed of which made ±1.0 m/year (Table 5.5).

138 glaciers were studied in the Enguri River basin. It turned out that 74.6% of them retreated. With high rate of retread were distinguished large valley glaciers. Eighteen glaciers in the Enguri River basin were stationary and seventeen glaciers advanced, mainly the hanging glaciers.

High speeds of retreating was characteristic to the glaciers on the northern slope of the of the Svaneti range. They were divided into separate glaciers during this period. Glaciers of the Dolra River basin suffered a strong degradation. Kvishi glacier has retreated by ∼500 m, Dolra glacier—by ∼750 m and the Baki glacier—by ∼1600 m. Almost all of the hanging types of glaciers of the Dolra river basin have advanced.

Over the period of 1946–1960, some of the valley glacier split (Zopkhito-Laboda, Lailchala, and Notsara, Kibesha, Denkara, and others). Before splitting the retreating speed of the valley glaciers was 40 m/year or more, and after splitting their retreating speed was about 20 m a year.

On the southern slope of the Caucasus 360 glaciers were surveyed with the help of the topographic maps, out of which 273 glaciers were retreating, 83 were in the stationary condition, and only 49 glaciers advanced (Table 5.5).

On the north slope of the Caucasus only the Tergi River basin, 45 glaciers were studied

(Table 5.5). Table clearly shows that the glaciers were retreating with the speed of 10.0 m/year. Such retreating rate was characteristic to the large glaciers such as Gergeti, Devdoraki, Abano, all three glaciers of Suatisi and others.

Many scientists have studied the glaciers in the Tergi River basin, but they say nothing in their works about the glaciers fluctuations in the years of 1946–1960. According to the data of Tsomaia (1960), the Devdoraki ice tongue retreated by 31.6 m (10.5 m/year) in the years of 1959–1962. Gergeti glacier retreated by 178.8 m (16.6 m/year in average) during the 1951–1963.

In Georgia we have fully studied 405 glaciers, out of which 64.4% have retreated, 20.5% were in the stationary state and only 12.1% of the glaciers advanced. Main the hanging types of glaciers have advanced. In the stationary state were the valley glaciers, which have the vast firns and the icefalls in the ice tongues. Glacier basin's morphological, morphometric, and climatic conditions identify the different speeds of the glacier retreat.

5.3 Dynamics of the Glaciers in 1960–1985

The well-known cartographic method of Glaciers variability study allows us to investigate the dynamics of the glaciers in the interval, when the cartographic materials were compiled. In the former Soviet Union cartographical materials (topographical maps) were renewed once in the 20–25 years, but the renewal had its negative sides as follows: during the map updating the attention was drawn not to the change of the relief surface, but to the renewal of the socioeconomic elements (such as roads, transmission lines, population statistical data, agricultural lands, etc.).

It was believed that in the conditions of the mountainous relief the dynamic development did not take place in the relief and its surface remained unchanged. Due to this, the content of the old maps was technically depicted in the new maps. We consider drawing up maps by such approach incorrect. It is necessary to reflect the earth's surface and the forms of geographical

landscapes distributed on it based on the making the new aerial photos. Such approach allows us to study the dynamics of the nival-glacial processes in the high mountainous areas not for individual objects, but also for the entire individual river basins or large orographical units as well. But as we have already noted, we are not able to do it.

Therefore, in order to study the dynamics of some of the glaciers, we used several different research methods, namely, R. Gobejishvili surveyed the Kirtisho, Adishi, Buba, Marukhi, Klichi, and Zopkhito glaciers by the repeated phototheodolite method; by means of labeling the observations were conducted on the following glaciers: Shkhara, Naumkvani, Chalaati, Kvishi, Dolra, and Tbilisa. The rest of the glaciers are investigated on the basis of aerial photo-materials of 1960–1973 (Gobejishvili 1995).

On the basis of the studies that were carried out, Table was compiled, which shows the variability of glaciers in 1960–1985 (Table 5.6).

Table analysis shows that the studied glaciers retreated with the different speeds, and there are the glaciers which maintained the stationary state against their background, or they were characterized by the small advances.

Let us consider some of the glaciers.

The Buba glacier is located in the Bubistskali River basin; morphologically it belongs to the compound-valley type of glacier. The Buba glacier is composed of three flows, which are combined below the ~3300 m and flows down in form of a common tongue. The glacier flows are separated from each other with the medial moraine. Glacier tongue flows movement velocity is nonuniform, these flows are different even by morphological signs. The Buba glacier's central flow was fed from the Karaugomi glacial plateau.

Variability of the Buba glacier tongue is studied by the repeated phototheodolite photographing method (Gobejishvili 1995). The first photographing was carried out in August 1971; the second photographing took place in August 1984. The interval between the photographing is 13 years.

Table 5.6 Changes of the glaciers in 1960–1985

Glacier no.	Glacier name	Name of the river basin	Change of the ice tongue Total ($\sim$ m)	$\sim$ m/year
22	Marukhi	Kodori	−10	−0.6
79	Klichi	–	−14	−0.8
80		–	−47	−3.1
81		–	0	0
82		–	0	0
119	Sakeni	–	−40	−2.8
136	Kharikhra	Enguri	−65	−4.3
206a	Ladevali	–	−1100	−55
206b	Tsalgmili	–	−125	−6.2
206c	Lakra	–	−450	−22.5
206	Kvishi	–	−1800	−110
209	Dolra	–	−250	−12.5
214	Chalaati	–	−52	−3.4
218	Lekhziri	–	−240	−17.1
223	Murkvami	–	−300	−21.4
240	Adishi	–	−10	−0.7
297	Didi Edena	Rioni	−37	−2.5
298	Patara Edena	–	−6	−0.4
300	Zopkhito	–	52	−3.5
302	Tsiskara	–	+6	+0.4
303	Laboda	–	+13	+0.9
318	Kirtisho	–	−110	−7.5
322	Khvargula	–	+12	+0.8
330	Buba	–	−50	−3.5

On the basis of the 1971 phototheodolite photographing the glacier tongue plan of 1:2000 scales has been done.

The model's scale was twice as large (1:4000) than the stereoautograph, with the help of which was drawn up the plans. The length of the basis to be photographed (phototheodolite) was 80 m on-site. Over the same heights (horizontal) of the topographic surfaces of the glaciers of 1971–1984 are depicted in the plan by the lines of different marks (Fig. 5.1).

The drawn plan shows the variability, which has undergone a glacier tongue in the interval among the potographings. According to the developed plan it is not difficult to characterize the variability of the ice tongue for the same point. A detailed grid method developed by G.A. Avsiuk was used to get the quantitative indicators for altitudinal variability in the same identical point of the glacier surface. The obtained data well characterize the variability of the topographic surface of the glacier in the years of 1971–1984 (Table 5.7).

The analysis of Table and plan compiled clearly showed the links that exist on the one hand between the glacier altitudinal distribution and thawing and on the other side, between the degree of glacier's topographical surface pollution and the surface thawing.

The analysis of the conducted researches shows that the Buba glacier tongue configuration has been changed drastically. For 13 years the

Fig. 5.1 The Buba glacier tongue plan drawn up in 1971–1984 according to the phototheodolite photographing materials: *1* The glacier boundary in 1971; *2* The glacier boundary in 1984; *3* The *horizontal line* on the surface of glacier of 1971; *4* The *horizontal line* on the surface of glacier of 1984; *5* Moraines on the glacier surface; *6–7* Micro-stade and stade moraines (Gobejishvili 1995)

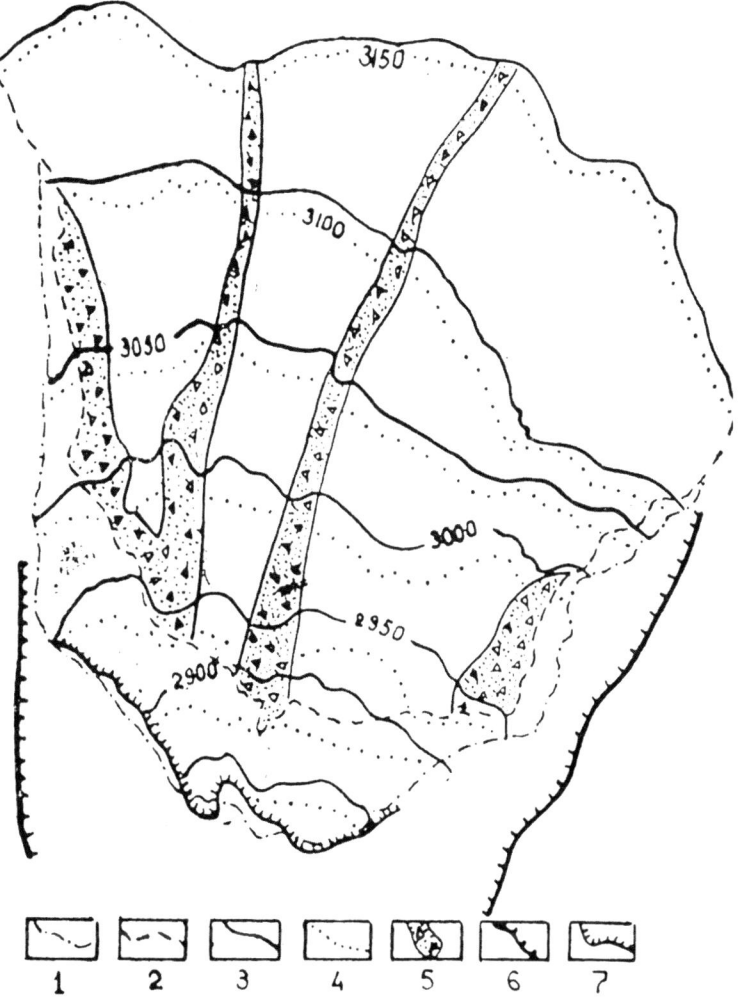

Table 5.7 The Buba glacier tongue thickness reduction according to the hypsometrical zones (1971–1984)

Hypsometrical zones (m)	Tongue thickness reduction (∼m)	Right flow (∼m)	Central flow (∼m)	Left flow (∼m)
2900–2950	19.7	13.7	21.9	23.5
2950–3000	15.2	18.8	13.5	13.4
3000–3050	13.2	14.2	12.6	12.8
3050–3100	9.7	10.1	9.2	9.8
3100–3150	6.8	8.4	6.6	5.6

glacier tongue retreated by ∼130–140 m, i.e., by ∼10–11 m per year. The glacier tongue's lateral sections were reduced by ∼30–40 m to the absolute height of 3250 m, more above the glacier tongue in the stationary state.

The microstade moraine formed in front of the glacier repeats the ice tongue configuration of the year of 1971. The measurements (according to the marks) conducted in the years of 1970–1973 and the description of the relief in front of the glacier showed that this microstade moraine was formed during this period (Gobejishvili 1995).

In the years 1960–1975 the valley's simple glaciers maintained the stationary condition. Their movement speed ranged as ±1 m/year. In front of such glaciers tongue almost everywhere the sharply developed microstade moraines are located.

Certain group of the glaciers is characterized by the intense retreat. These are the compound-valley glaciers, which were split or a new stream was separated from them.

The exclusive retreat suffered the glaciers of the Kvishi River basin. The **Kvishi glacier** was consisted of four main flows according to the 1960 aerial images, and it was a compound-valley glacier. Glaciers common tongue was of the eastern exposition. Glacier area was 19.10 km². The ice tongue ended at the height of 2415 m above sea level. The common ice tongue was covered with the surface moraines and it seemed as if the flows of the individual glaciers were not joining each other. The field observations showed that the glacier tongue was not a motionless buried ice. It was active due to glaciers tongue pressure. The situation of the glaciers of that period is well described by our preceding researchers (Kovalev 1961; Khazaradze 1971; Gobejishvili 1995).

In 1960, the right stream of the glacier was flowing from the Ladevali peak; this branch, which is now an independent glacier, is called Ladevali. Even today the glacier has a vast firn valley and well-expressed ice tongue; after flowing out from the firn it develops the icefall. The left flow, which has a vast firn valley, joined the Ladevali glacier in the form of narrow tongue. Today this flow is alone and separated from Ladevali by the ledge. We name the glacier as Tsalgmili, because it begins on the slope of the Tsalgmili peak. The Ladevali glacier tongue is polluted with the materials brought by the snow avalanches. The medial moraine stretches along the glacier tongue surface. The glacier tongue sinks under the powerful medial moraine of the Kvishi glacier. The Tsalgmili glacier consists of two flows; above the ledge the medial moraine is developed. The lateral stade moraine is developed

in the right side of the Ladevali glacier, which is deformed in the last section of the ice tongue due to the gravitational processes.

The Lakra glacier is a valley type of glacier with the eastern exposition. It consists of two branches. It flows from the firn by icefall and the tongue is cracked. Ice passes from coming out and sign language. Along its entire length the medial moraine is stretched. The stade moraine is only on the left side and even it is deformed.

Kvishi glacier is the most powerful among the flows described above. It is fed from the vast firn valley. After leaving the firn the ice tongue is characterized by a steep inclination and is dissected with the transverse and longitudinal cracks. The glacier tongue is still covered by the loose materials (Fig. 3.28). Earlier it was covered by the more powerful loose material and was a connector for all of the flows. Today it is ended by an independent tongue. The Kvishi glacier firn is two-layered and among them the icefall is developed.

As we have already mentioned, four glaciers were represented in one tongue. In the years of 1965–1975 they were divided and they now represent the separate glaciers. During the period of degradation these glaciers retreated in different speeds. Kvishi glacier retreated very fast—by ∼90 m/year. In total—∼1800 m. The Ladevali glacier retreated by ∼1100 m (∼55 m/year) and the Tsalgmili glacier—by ∼125 m (6.2 m/year). In 1975 this glacier almost no longer connected to the common ice tongue (Kovalev 1961). Kvishi and Tsalgmili glaciers retreated by ∼3–5 m/year in the period of 1977–1982.

The Lekhziri, Tviberi, Kvitlodi, and Seri glaciers retreated quickly. These are the glaciers, the tongues of which are covered with the powerful loose materials.

The employees (Tsomaia and Drobishev 1970) of the Hydro-meteorological Service and Hydrological and Meteorological Institutes were monitoring the fluctuation of the glaciers. Their observation materials show that glaciers are moving in different ways (Table 5.8).

Analysis of table shows well that the eastern Georgia's glaciers advanced by ∼0.5–5.8 m/year speed. Over the period of 1973–1978, only the

Table 5.8 The variation values of the glaciers of Georgia according to the different years

Glacier name	Observation period	Change of the tongue's end	
		Total (m)	m/year
Devdoraki	1963–1972	+28.8	+3.2
		+25.7	+5.1
Gergeti	1973–1978	+355	+3.9
		−5.9	−1.2
Suatisi	1963–1972	+17.8	+3.6
		+29.2	+5.8
Kirtisho	1973–1978	−46.6	−7.8
		−53.8	−13.5
Chalaati	1963–1972	−40.2	−6.7
		−27.3	−6.8
Koruldashi	1973–1978	−32.8	−5.5
		−14.8	−3.7
Laghztsiti	1966–1972	+4.0	+0.8
		+2.0	+0.5

Gergeti glacier retreated. The western Georgia's glaciers had retreated annually in the years of 1973–1978. Against the background of common retreat of 1966–1973 the glaciers of Kirtisho and Chalaati advanced by ∼1.0–2.0 m in 1970–1971, and the Koruldashi glacier has retreated only by ∼0.6 m. In the years of 1970–1971 the eastern Georgia's glaciers advanced as well: the Gergeti glacier—by ∼4.8 m, the Devdoraki glacier—by ∼18.0 m and the Suatisi glacier—by ∼7.2 m. In the years of 1970–1971 the similar activity was recorded by R. Gobejishvili in the glaciers of Buba, Tbilisa, Laboda, and Adishi. In this period the microstade moraines occurred in front of the glaciers tongues.

A study conducted in 1975–1985 (R. Gobejishvili), in particular in 1984–1985 the stamps repeated examination showed, that the variability of the glaciers is different. At this time, some of the glaciers advanced, some retreated, but this variability is small and they are considered stationary glaciers.

In 1974–1985 the Adishi glacier advanced by ∼1.0 m, the Shkhara glacier—by ∼5 m in 1978–1985, and the Naumkvani glacier retreated by ∼32.0 m in the same period.

The glaciers of the Rioni River basin—Edena, Khvargula, Laboda, and Tsiskara advanced by ∼5–15 m during 1970–1985, and the Kirtisho glacier has retreated in average by ∼80–100 m. The glaciers of Zopkhito, Tbilisa, Didi Edena, and Brili have retreaded by ∼50–90 m.

As the researches show, the glaciers variability is not of a single-sign and is not of a simple nature, which related to the location at the different heights above sea level, existence of the surface moraines, glaciers exposition, their area, climatic conditions, etc.

5.4   Dynamics of the Glaciers in 1960–2014

Since the end of 1950s until the middle of 1970s, the glaciers were in a quasi-stationary state in most of the mountain areas in Eurasia (Dyurgerov 2005). Now, the glaciers degrade in all mountain areas of Eurasia (Khromova et al. 2014). In the Caucasus, particularly, in Georgia, the glacial processes also prove it. Generally, in most cases as a result of glaciers melting their morphological types, exposition and elevation above sea level change, their area reduces and their number increases. All these processes fit within the norms in case of Georgia, except of the increasing in the area in parallel with the reduction in the area of the glaciers.

Glaciological and geomorphological researches conducted in the Caucasus by us has revealed variability of morphological types of glaciers, their exposition, height of glacier tongues, snow and firn lines during 1960-2014. As for the area and the number of the glaciers, there is a decrease here in both cases.

Recently the full statistical and descriptive information about the glaciers of Georgia is presented by R. Gobejishvili in 1989 (Glaciers of Georgia 1989), which is based on 1:25,000 and 1:50,000 scales topographic maps compiled in 1960s and the materials obtained by the processing of the aerial images of that period. According to him there were 786 glaciers in Georgia in 1960 with the total area of 563.70 ± 11.31 km^2 (Table 5.9).

Almost half a century later, based on the modern methods, new aerial images (Landsat L8 and ASTER 2014) and field expeditions we obtained the latest information about the glaciers of Georgia. Our research has found that for the period 1960–2014, the number and area of the glaciers were reduced, respectively, by 19.0 ± 5.0 and $36.0 \pm 2.0\%$ (Tielidze 2016) (Table 5.9). Such a sharp reduction in the number of glaciers is caused by the fact that in Georgia— 1960–1970s of the twentieth century—most of

the glaciers were small cirque-type of glaciers, which have disappeared completely in the last half century (Tielidze 2014). Change of glaciers according to separate parts of the Caucasus range and river basins are following.

The Western Caucasus. The Bzipi River gorge is the westernmost basin of the territory of Georgia, where the contemporary glaciers are represented (Tielidze 2014). Except of Bzipi, the glaciers are represented in the basins of the rivers of Kelasuri and Kodori within the Western Caucasus.

By the data of 1960 maps there were 18 glaciers with the area of 9.90 ± 0.20 km^2. According to the aerial images of 2014 the number of the glaciers is 18, while the area 3.99 ± 0.13 km^2 (Table 5.9). The Bzipi River basin is characterized by the cirque glaciers of small size of about 0.5 km^2.

By the data of 1960 there was only one glacier with the area of 0.26 ± 0.015 km^2 in the Kelasuri basin. According to the data of 2014 the number of the glaciers is 1, while the area 0.11 ± 0.005 km^2 (Table 5.9).

The major center of the modern glaciation on the southern slope of the Western Caucasus is

Table 5.9 The change in the area and number of the glaciers of Georgia in 1960–2014 according to the individual river basins

Basin name	Topographic maps of 1960, based on the 1955–1960 aerial images			Landsat and ASTER Imagery (2014)		
	Number	Area (km^2)	Uncertainty (%)	Number	Area (km^2)	Uncertainty (%)
Bzipi	18	9.90 ± 0.20	±2.07	18	3.99 ± 0.13	±3.25
Kelasuri	1	0.26 ± 0.015	±5.76	1	0.11 ± 0.005	±4.54
Kodori	160	63.73 ± 1.63	±2.55	145	40.06 ± 1.29	±3.22
Enguri	299	323.70 ± 5.72	±1.76	269	223.39 ± 4.6	±2.05
Khobistskali	16	1.12 ± 0.07	±6.25	9	0.46 ± 0.03	±6.52
Rioni	112	76.77 ± 1.66	±2.16	97	46.65 ± 1.15	±2.47
Liakhvi	16	4.27 ± 0.13	±3.04	10	1.82 ± 0.07	±3.84
Aragvi	3	0.88 ± 0.03	±3.40	1	0.31 ± 0.015	±4.83
Tergi	99	67.01 ± 1.33	±1.99	58	35.56 ± 0.8	±2.24
Asa	9	2.59 ± 0.085	±3.28	3	0.54 ± 0.025	±4.62
Arghuni	17	2.92 ± 0.12	±4.10	6	0.43 ± 0.025	±5.81
Pirikita Alazani	36	10.48 ± 0.32	±3.10	20	2.42 ± 0.11	±4.54
Total	786	563.70 ± 11.31	±2.00	637	355.80 ± 8.25	±2.32

located in the Kodori River basin, which extends from the Marukhi pass to the Dalari pass. The height of the peaks located there exceeds 3800–4000 m. By the data of 1960 there were 160 glaciers with the area of 63.73 ± 1.63 km^2. According to the data of 2014 there are 145 glaciers in this basin with the total area of 40.06 ± 1.29 km^2 (Table 5.9).

In total, in the Western Caucasus glaciers area decreased by $40.23 \pm 2.87\%$ or 29.73 ± 1.62 km^2 from 73.89 ± 1.85 to 44.16 ± 1.43 km^2 in the years of 1960–2014.

The Central Caucasus section is distinguished by the highest relief in the territory of Georgia, the height of the peaks located there exceeds 4500–5000 m. There are the river basins within the Central Caucasus such as the Enguri, Khobistskali, Rioni, and Liakhvi.

Enguri River basin is the largest in Georgia according to the number and area of the contemporary glaciers. It exceeds all other basins taken together. The largest glaciers of Georgia such as Lekhziri, southern and northern Tsaneri, and others can be found here (Tielidze 2014). According to the data of 1960, there were 299 glaciers with the area of 323.70 ± 5.72 km^2, and by the data of 2014 there are 269 glaciers in this basin with the total area of 223.39 ± 4.6 km^2 (Table 5.9).

By the data of 1960 there were 16 glaciers with the total area of 1.12 ± 0.07 km^2 in the Khobistskali River basin. According to the data of 2014 there are nine glaciers in this basin with the area of 0.46 ± 0.03 km^2 (Table 5.9).

Another important center of the modern glaciation in Georgia is the Rioni River basin. According to the data of 1960, the number of the glaciers was 112 with the total area of 76.77 ± 1.66 km^2 in the Rioni River basin. By the data of 2014, there are 97 glaciers with the total area of 46.65 ± 1.15 km^2 (Table 5.9). The largest glacier in the Rioni River basin is Kirtisho with the area of 4.41 ± 0.07 km^2.

By relatively low hypsometrical, location is distinguished the Liakhvi River basin, which is the easternmost basin of the Central Caucasus. There were 16 glaciers in the Liakhvi River basin

with the total area of 4.27 ± 0.13 km^2 according to the data of 1960. According to the data of 2014, there are 10 glaciers in the Liakhvi River basin with the total area of 1.82 ± 0.07 km^2 (Table 5.9).

In total, the glaciers area decreased by $32.90 \pm 2.01\%$ or 133.54 ± 4.87 km^2 from 405.86 ± 7.59 to 272.32 ± 5.86 km^2 in the period of 1960–2014 in the Central Caucasus.

In Georgia **the Eastern Caucasus** is represented both by southern and northern slopes. The basins of the rivers such as Aragvi, Tergi, Asa, Arghuni, and Pirikita Alazani are located there. Most of these river basins are distributed on the northern slopes of the Caucasus.

The westernmost basin of the Eastern Caucasus, where the contemporary glaciers are present, is the Aragvi River basin. According to the data of 1960, there were three glaciers in the Aragvi River basin with the total area of 0.88 ± 0.03 km^2. By the data of 2014, the only one glacier (Abudelauri) is remained in the basin with the area of 0.31 ± 0.015 km^2 (Table 5.9).

The Tergi River basin is a main glaciation center of the Eastern Caucasus. Some of the peaks' heights exceed 5000 m here (Mkinvartsveri/ Kazbegi 5033 m). According to the number of glaciers the Tergi River basin is in the fourth place after Enguri, Kodori, and Rioni and its share is $\sim 9.1\%$ in the total number of the glaciers of Georgia. By the data of 1960, there were 99 glaciers with the total area of 67.01 ± 1.33 km^2. According to the data of 2014, there are 58 glaciers with the total area of 35.56 ± 0.8 km^2 (Table 5.9).

The Asa River basin is located on the northern slope of the Greater Caucasus. Its name is Arkhotistskali in the territory of Georgia. By the data of 1960, there were nine glaciers in this basin with the total area of 2.59 ± 0.085 km^2. According to the data of 2014, there are three glaciers with a total area of 0.54 ± 0.025 km^2 (Table 5.9).

The Arghuni River basin is located on the northern slope of the Greater Caucasus and it has the meridional direction. By the data of 1960, there were 17 glaciers with the total area of

2.92 ± 0.12 km^2. According to the data of 2014, there are only six glaciers with the total area of 0.43 ± 0.025 km^2 (Table 5.9).

Pirikita Alazani River basin is located on the northern slope of the Greater Caucasus and is of latitudinal direction. Here the individual peaks' height is over 3800–4000 m. By the data of 1960, there were 36 glaciers with the total area 10.48 ± 0.32 km^2. And according to the data of 2014, there are 20 glaciers in this basin with the total area of 2.42 ± 0.11 km^2 (Table 5.9).

In total, the glaciers area decreased by $53.19 \pm 2.38\%$ or 44.62 ± 1.44 km^2 from 83.88 ± 1.90 to 39.26 ± 0.98 km^2 in 1960–2014 in the Eastern Caucasus.

In order to have a clear picture of the variation of number and area of the glaciers in the Caucasus during the last half century, we distinguished the Chalaati and Zopkhito glaciers as a model. By using the modern methods, latest aerial images and different computer softwares we conducted the study of the glaciers regarding the climatic elements. Our research found that both glaciers have undergone substantial changes during the modern record (1960–2013).

Over the study period, Chalaati Glacier area decreased by 4.219 km^2 ($\sim 33\%$), equivalent to ~ 0.079 km^2/year (Table 5.10). Rates of area loss have been variable, with the fastest rate (~ 0.159 km^2/year) occurring between 1987 and 2000. During the next 13-year-period (2001–2013), the rate of area loss decreased to ~ 0.058 km^2/year. The most noticeable change on Chalaati Glacier is the separation of two small tributaries from the main channel of the glacier, which occurred between 1987 and 2000. These tributaries originate on the slopes of Chatini Peak and their disconnection represented a loss of 1.900 km^2 (Fig. 5.2).

In addition to area changes, we also map linear retreat of the terminus since 1960 (Fig. 5.3). In contrast to the area loss, the fastest rates of snout retreat occurred during the most recent epoch (2001–2013).

We mapped the annual snowline position from satellite images, all of which were collected close to the end of the ablation season in August, for comparison with snow line positions in 1960 mapped by Gobejishvili (1995). The firn line on Chalaati Glacier was located at 3155 m asl in 2013. The firn basin extends to 3800 m and covered an area of 4.161 km^2. Between 1960 and 2013, the elevation of the firn line has risen by 35 m, resulting in a decrease in the accumulation zone of ~ 0.800 km^2 over the 53 year period. Using the measured geometries, we calculate the accumulation-area ratio (AAR) (Table 5.11). Chalaati Glacier's total area has decreased at a faster rate than the accumulation area, so the AAR has actually increased over the study period (Table 5.11).

Over the study period, the area of Zopkhito Glacier decreased from 2.886 to 2.429 km^2 ($\sim 16\%$), equivalent to a rate of ~ 0.008 km^2/year (Table 5.10). The rate of area loss was fastest during the most recent part of the record (~ 0.081 km^2/year between 2000 and 2013). Most of the area loss occurred in the lower parts of the glacier, where there was a steady rise in the altitude of Zopkhito Glacier's terminus, from

Table 5.10 Morphometric parameters of the Chalaati and Zopkhito glaciers in 1960–2013

Years	Tongue height above sea level (m)		Area (km^2)					Length (km)	
	Chalaati	Zopkhito	Chalaati		Zopkhito			Chalaati	Zopkhito
1960	1800	2435	12.814 ± 0.353	Loss%	2.886 ± 0.085	Loss%		7.62	4.33
1986	1890	2525	11.598 ± 0.365	9.48	2.669 ± 0.081	7.51		7.12	3.95
2000	1930	2550	9.359 ± 0.278	19.30	2.575 ± 0.079	3.52		7.05	3.84
2013	1960	2605	8.595 ± 0.279	8.16	2.429 ± 0.071	5.66		6.81	3.58

Fig. 5.2 Two small separated tributaries (**a** and **b**) from the main channel of the Chalaati glacier (*photo by* L. Tielidze)

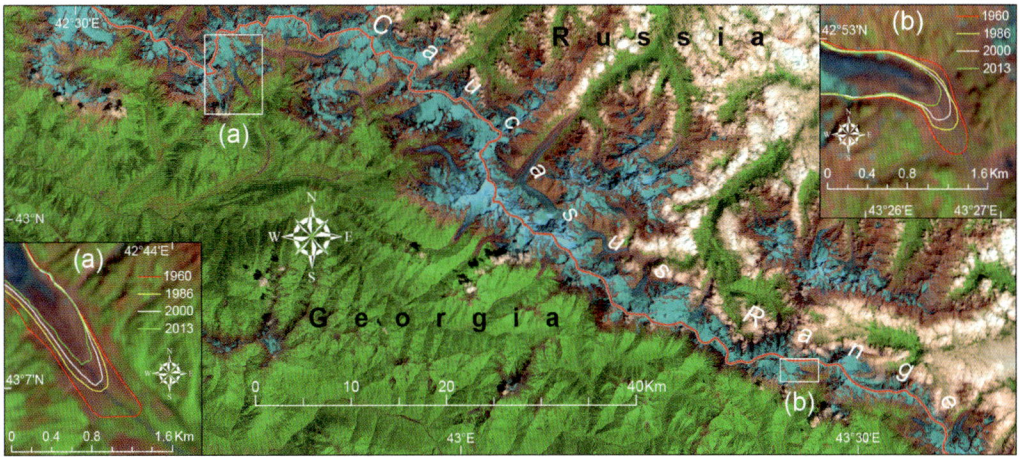

Fig. 5.3 Chalaati (**a**) and Zopkhito (**b**) Glaciers retreat in 1960–1986–2000–2013 (Landsat L8 OLI 23.08.2013)

Table 5.11 AAR for Chalaati and Zopkhito glaciers, 1960–2013

Years	Accumulation zone (firn) area (km^2)		Total area (km^2)		AAR	AAR
	Chalaati	Zopkhito	Chalaati	Zopkhito	Chalaati	Zopkhito
1960	4.961	1.847	12.814	2.886	0.387	0.639
1986	4.646	1.739	11.598	2.669	0.400	0.651
2000	4.459	1.688	9.359	2.575	0.476	0.655
2013	4.161	1.626	8.595	2.429	0.484	0.669

2435 m asl (1960) to 2605 m asl (2013). This led to a ~0.752 km reduction in the length of the glacier since 1960.

In 2013 Zopkhito Glacier's snow line was located at 3080 m asl and encompassed a 1.847 km^2 firn basin extending to 3800 m. There was a 30 m rise in the elevation of the snow line between 1960 and 2013, resulting in a decrease in the accumulation zone of ~0.221 km^2 over the 53 year period. Using the measured geometries, we calculate the accumulation-area ratio (AAR) which shows Zopkhito Glacier's total area decreased at a faster rate than the accumulation area, leading to an increasing AAR over the study period (Table 5.11).

Our results show that in our sample two glaciers on the southern slope of the Central Caucasus are experiencing sustained retreat. We examine these changes in the context of regional air temperature conditions. Meteorological measurements only exist for occasional short (2–3 months) periods at Chalaati (summer 2011) and Zopkhito glaciers (summers 2009), but detailed records are kept at the weather station in the settlement of Mestia, which is located ~7 km down the valley from the terminus of Chalaati Glacier (~63 km from Zopkhito Glacier), at an elevation of 1440 m. Observations are available for the period 1961–2013.

The mean annual temperatures at both glaciers are below freezing for the entire record. Zopkhito Glacier is colder than Chalaati, as would be expected from its higher elevation terminus (~700 m higher than Chalaati) (Table 5.10). In general, the warmest temperatures occur in July (Fig. 5.4), and the melt season (defined as temperatures above 0 °C) lasts an average of 184 days at Chalaati and 145 days at Zopkhito. There has been an increase in the length of the melt season at both glaciers (Fig. 5.5). The increasing length of the melt season is consistent with a general trend of warming air temperatures anomalies over the period 1960–2013 (Fig. 5.6). The warmest temperatures have occurred since the mid-1990 s coinciding with the fastest observed rates of glacier total area reduction.

Field observations at both glaciers of exposed stake heights allow us to examine the role of air temperature on ablation. On the basis of the derived lapse rates, we can use the Mestia temperature record to produce a "local" temperature record for each stake location and compute the cumulative positive degree days (PDD) for each site (cf. Hock 1999).

The observations for the Chalaati Glacier were made in summer 2011 at an elevation of 2040 m asl. The analysis yields an ablation rate of ~0.6 cm/PDD in July and ~0.4 cm/PDD in August. Repeating the analysis for Zopkhito Glacier using observations from 2009 for a stake at 2700 m asl yields an ablation rate of ~0.5 cm/PDD in July and ~0.3 cm/PDD in August. This results from falling air temperatures responding to lower solar angle toward the end of summer.

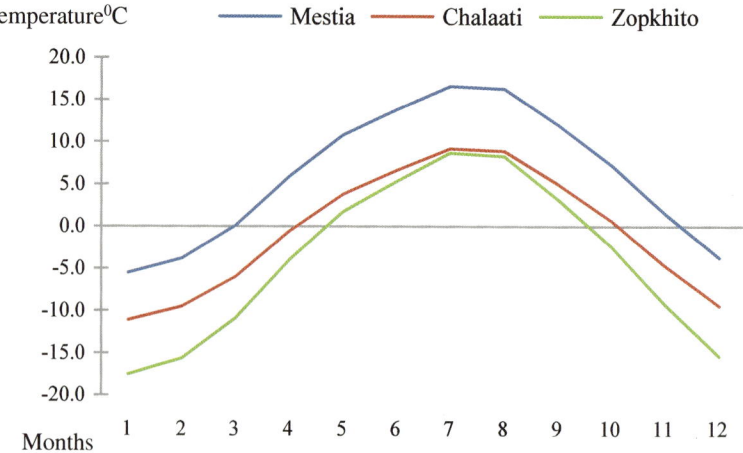

Fig. 5.4 Average monthly air temperatures (1960–2013) at Mestia weather station and Chalaati and Zopkhito glaciers

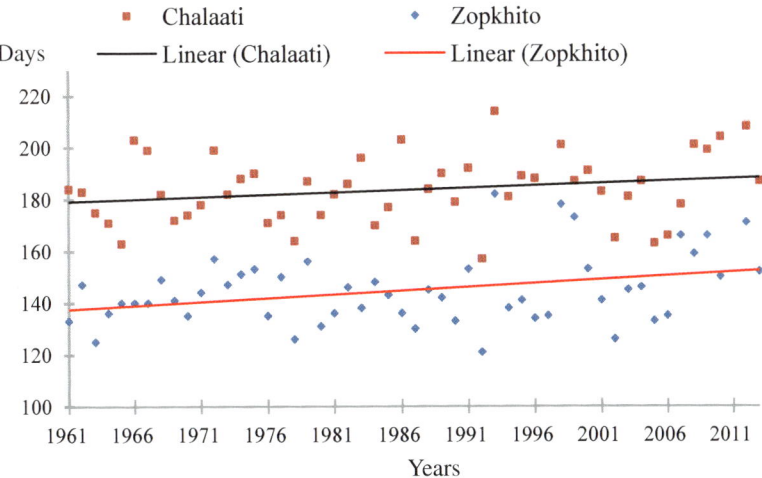

Fig. 5.5 Melt season durations at Chalaati and Zopkhito glaciers over the period 1960–2013 and linear trends

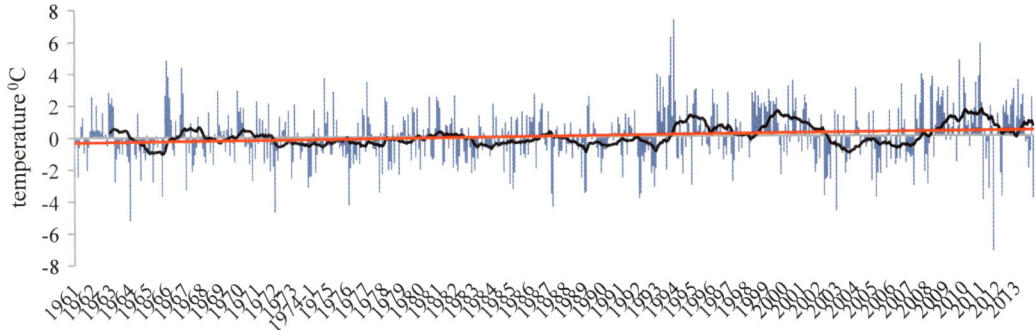

Fig. 5.6 Time series of monthy air temperature anomalies at Mestia weather station with respect to the 1960–2013 average. A 24 month smoothed anomaly is shown by the *thick blue line*. The *red line* is the trend showing a modest increase in warm anomalies with time

Our results are consistent with other studies of glacier changes in the Central Caucasus Mountains (e.g., Shahgedanova et al. 2014), although most previous studies have focused on the north-facing slope (Russian side). Rates of area loss for Chalaati and Zopkhito glaciers are on the upper range of losses reported by Shahgendanova et al. (2014) who found generally lower rates of glacier wastage for north-facing glaciers. Our results show that glaciers with aspects conducive to increased solar radiation are more sensitive to temperature-driven ice loss. One consequence of this result is that Georgian glaciers are at higher risk of disappearance than north-facing glaciers in Russia.

Glaciers on the southern slopes of the Caucasus Mountains are expected to continue their retreat as regional air temperatures get warmer. This deglaciation will have important consequences for the management of water resources for agriculture and hydropower production in Georgia.

5.5 Valley Glaciers Reduction After the Little Ice Age Maximum

In glaciological sense, the duration of Little Ice Age (LIA) is comparable to a period, when glaciers were larger than before or after that time

(Holzhauser 1983). The end of the LIA Maximum in different mountainous systems of the world is recorded about 150–200 years ago. The LIA was a global climatic deterioration at a scale of hundreds of years, which has strongly influenced on natural environment and human civilization. In polar and high mountain areas, glaciers advanced and achieved their maximum postglacial positions during this period. The strongest impact of cooling was noticeable in the North Atlantic region: in northern latitudes of Europe, Asia, and North America (Hunt 2006; Zasandi 2007). From the beginning of about nineteenth century (1820–1850) the climate conditions have changed over the Earth's surface and the warming began, which is still going on. This process was followed by the retreat of glaciers in almost all mountainous system (with some local exceptions).

One of the main centers of the Eurasian glaciation during the LIA was the Caucasus Mountain range, where the two main glacial advances phases within the LIA have been recognized, the first in the second half of the thirteenth to the early fourteenth centuries, and the second in the early seventeenth to first half of the nineteenth centuries (Grove 1988).

Many present researchers also state that according to lichenometry in the Caucasus the second phase LIA Maximum really occurred between the end of the seventeenth and the first half of the nineteenth century (Kotlyakov and Krenke 1981; Grove 1988; Solomina 2000). The same prove the works of some of the Georgian scientists, that the LIA reached its maximum in the Caucasus in the beginning of the nineteenth century ($\sim$1820) (Gobejishvili 1995; Tielidze 2014). Besides the fact that the certain researches have been conducted in the Caucasus on LIA, the basic information is still available on the glaciers of the Russian Caucasus, and the information on the glaciers of the Georgian Caucasus of those times is not actually printed yet. Accordingly, we believe that the result of our research is an important information in the case of the study the LIA even more thoroughly.

In this chapter we present the results of the conducted researches in the different glaciers (in total, in the 24 glaciers) of the Western, Central, and Eastern Caucasus. With the help of the LIA's moraines, which are well preserved in the main glacial valleys, are reconstructed the parameters of the old glaciers, such as the area, width, retreat, and hypsometrical location of the ice tongues.

We have investigated almost all of the glacial basins in the Georgian Caucasus. Especially in detail, we studied the basins with the valley type of glaciers. We took pictures of the glacier tongues and moraines and conducted the survey by the GPS device. We also used old topographic maps of different years.

Besides the field works and topographical maps, it is very important to use the aerial images for study the LIA. Some researchers, who had conducted studies before us, mention that the LIA's moraines can be sharply seen in the aerial images of the Caucasus (Solomina 2000), which are often used as a reference point to estimate the magnitude of past glacier variations (Furrer et al. 1987; Grove 1988; Solomina 2000). Consequently, during the study of glaciers for its reconstruction purposes we use the modern aerial images and the remote sensing method. Landsat L8 OLI/TIRS (Operational Land Imager and Thermal Infrared Sensor), with 30 m horizontal resolution available since 2013, and Advanced Spaceborne Thermal Emission and Reflection Radiometer (ASTER) imagery with 15 m resolution available since 2000, together with the old topographical maps allow us to identify the change in the area and terminus of the glaciers in the last $\sim$2 century by a minimum error. All aerial images (Landsat and ASTER) are taken in 2014 in the cloudless weather in August–September, when the glaciers and their adjacent territories were free of the seasonal snow and the LIA's moraines can be seen well.

To have a minimum error, we also used the Google Earth software, which shows very clearly the LIA's moraines and a trace of the old glaciation. The moraines surveyed by us coincides to the moraines showed in the Google Earth image with the minimum error ($\pm$1 m horizontally).

The analysis of the mentioned work shows that the glaciers located in the Caucasus have been reduced greatly during the last ~200 years. The indicators of the glaciers reduction in length, area, and capacity differ from each other. Three groups of the glaciers can be distinguished according to the reduction rate: the first group includes the glaciers, the tongues of which have retreated by more than ~2.0 km, the second group includes the glaciers with the tongues reduced by ~1.5–2.0 km, and the third group includes the glaciers, the retreating indicators of which is less than ~1.5 km.

Compound-valley glaciers are mainly included **in the first group**; classical example of the glaciers of this group is the **Tviberi glacier**, which is located in the river basin with the same name (the Tviberi River is a mouth of the Mulkhura River). It is located on the southern slope of the Caucasus. In spite of this, the peaks located there are not distinguished by the high altitude. The highest peak is of 4250 m and is lower than the peaks located next to it by 600–800 m. Morphological conditions of the relief play an important role in the forming of the powerful glaciations in the basin together with the climatic conditions. The Tviberi River basin is surrounded by the quite high ranges. Glacial forms are deeply sat in the relief, built by the crystal rocks and create the favorable conditions for accumulation of snow-icy cover. From the beginning of nineteenth century to the second half of twentieth century the Tviberi River basin's glaciers were combined and they formed the compound-valley glacier. In ~1820 their joint tongue was flowing down to the ~2010 m above sea level, for that time the glacier area was ~53.3 km^2 (Fig. 5.7). Tviberi glacier was the largest in area in Georgia for that time. Identification of the glacier maximum borders during the LIA is not difficult, because the glacial morphosculptural forms (moraines, ram forheads) have been remained in the relief. They are created due to the glaciers advancing during the LIA maximum (Fig. 5.7).

As we already have mentioned, from the year of ~1820 the glaciers of the Caucasus are retreating. In the topographical map of nineteenth

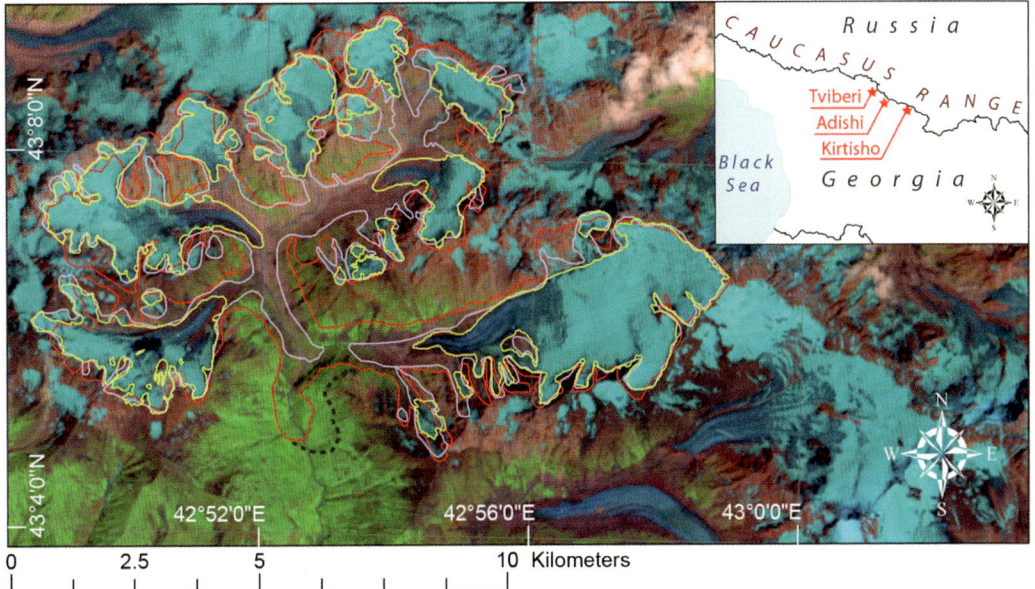

Fig. 5.7 Landsat L8 image of 2014. Tviberi glacier reduction from the LIA Maximum (~1820) through the year of 2014. *Black dash line* indicates the LIA Maximum moraine; the *red dye* shows the Tviberi glacier contour by the topographical map of 1887; the *violet dye* shows the contours of the glaciers of the Tviberi basin by the topographical map of 1960; the *yellow dye* shows the contours of the glaciers of the Tviberi basin by Landsat image of 2014

century (1887) the Tviberi glacier is presented as a united system. Its area was 49.02 ± 0.44 km^2 (Fig. 5.7). For that time its ice tongue was ended at a height of ~ 2030 m above sea level. Decoding of the aerial images of 1959–1960 and the analysis of the topographical maps showed that the Tviberi glacier has been greatly changed. Glacier area was decreased by ~ 24.3 km^2. The greatest branch—Kvitlodi was separated from the glacier's left side, which became an independent glacier. The Kvitlodi glacier tongue was located ~ 300–400 m away from the Tviberi glacier and ended at a height of ~ 2290 m. Five glaciers of small size were separated from the Tviberi glacier system as well (Fig. 5.7).

In 1960 the Tviberi glacier tongue was ended at a height of ~ 2140 m. The surface of its tongue was covered by the weathered material of ~ 0.5–0.8 m thick, which greatly reduced the surface ablation. A clean surface of the glacier was melting relatively greatly in the area of the joint tongue than the surface covered by the weathered material. Because of that the clean section of the glacier was located ~ 10–15 m lower. The area of the Tviberi glacier was 24.72 ± 0.35 km^2 in 1960 (Gobejishvili 1989).

Decoding of the nineteenth and twentieth centuries' topographical maps and aerial images of 2014 showed that due to the retreat the compound-valley glacier of Tviberi was divided into the following valley types of glaciers: Seri, Asmashi, Toti, Iriti, Lichati, Laskhedari, and Dzinali. The glacier tongues are ~ 1.0 km away from each other. The Asmashi glacier is the largest in area and length; its tongue ends at a height of 2540 m. The sides of the ice tongue surface are covered by the ~ 0.5–1.0 m thick weathered material and the ~ 200 m wide clean ice flow is intruded among them in the form of a wedge. Its surface is lower than the sides by ~ 5–10 m.

From the LIA Maximum to till 2014 the Tviberi glacier tongue retreated by ~ 4.4 km. Such a sharp change of the tongue can be observed with the help of the old and modern images as well (Fig. 5.8). Such a rate of retreat was characteristic to the Kvishi glacier in Svaneti, but after its destruction the rates of retreat of single glaciers have reduced sharply. Same process is in the Tviberi basin glaciers as well.

Simple-valley glaciers belong to the **second group,** which retreated by ~ 1.5–2.0 km after the LIA Maximum. For this kind of glaciers the well-expressed firn valley and ice tongue is characteristic. A classic example of this type of glaciers is the **Kirtisho glacier**, which is located in the Chveshura River basin in the Racha Caucasus. It is a simple-valley type of glacier with the north-western exposition. From 1937 the glacier observation started (Tsereteli 1959).

The moraine of the LIA Maximum has been remained on the both slopes of the gorge in front of the glacier (especially on the left slope) remained LIA's maximum Morena. By using the program Google Earth, old topographic maps (1887, 1960), data of Tsereteli (1959) and GPS

Fig. 5.8 Reduction of the Tviberi glacier in 1884 (*photo by* L. Tielidze)—2011 (*photo by* M.V. Dechy)

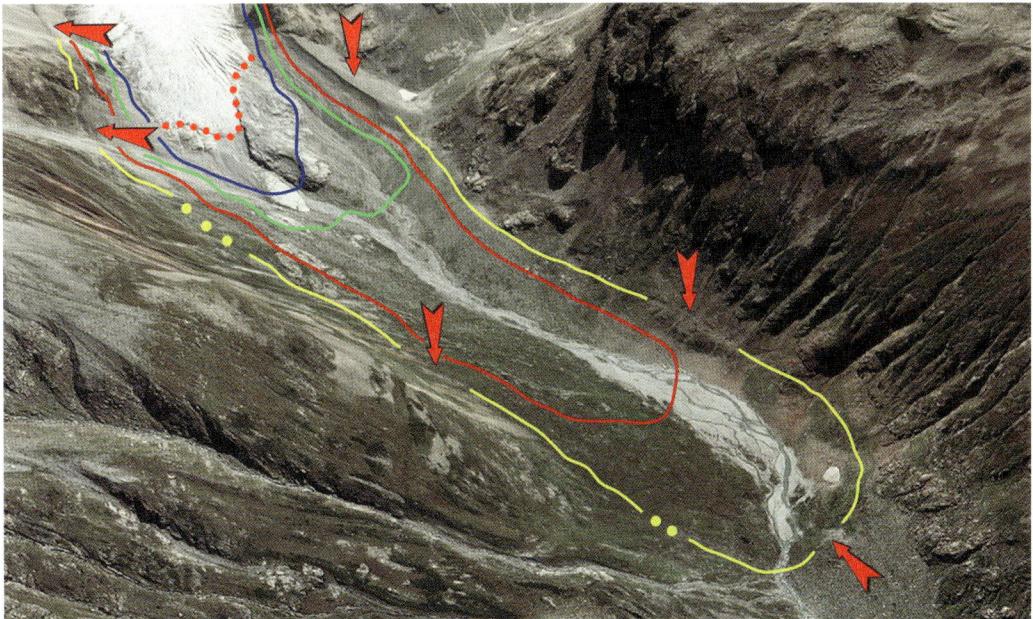

Fig. 5.9 Google Earth image 2013. Kirtisho glacier reduction from the LIA Maximum to 2014. *Red arrows* indicate the LIA Maximum moraines; *yellow dots* indicate the fragments of the LIA Maximum surveyed by GPS; *yellow line* indicates the reconstructed contour of the Kirtisho glacier in LIA Maximum; *red line* marks the glacier contour according to the topographical map of 1887; *green line* marks the glacier contour according to the 1937 (by the data of Tsereteli 1959); *Ice blue lines* mark the glacier contour according to the topographical map of 1960; *red dots* indicate the end of the glacier tongue in 2014 obtained by GPS surveying

data of 2014, we reconstructed the Kirtisho glacier reduction during the last two centuries (Fig. 5.9).

As it was identified, the Kirtisho glacier area was ∼6.6 km² for ∼1820 and its ice tongue descended to the ∼2325 m above sea level. In 1820–1887 (according to the topographical map of 1887) Kitrisho tongue retreated by ∼240 m, while hypsometrically it shifted to ∼2335 m above sea level. In 1887–1937 years (according to the data of 1937 by D. Tsereteli) the glacier retreated even by ∼740 m, while hypsometrically it shifted to ∼2380 m above sea level. In 1937–1960 (in 1960, according to the topographical map) the glacier retreated by ∼300 m, while hypsometrically it shifted to ∼2420 m above sea level. In 1960–2014 the glacier retreated by ∼460 m. Now the glacier tongue ends at the height of 2660 m above sea level. In total, the glacier retreat in 1820–2014 made ∼1740 m, while its area decreased from ∼6.6 to ∼4.4 km².

The third group unites the simple-valley glaciers, the ice tongues of which retreated less than ∼1.5 km after the LIA Maximum. The classical example of this group of glaciers is the **Adishi glacier**, which is located in the Enguri River basin. The glacier has an expanded firn, which is located at a height of ∼3700 m. The firn is fed by several flows. After leaving the firn the glacier develops a powerful icefall of the height of ∼1000–1500 m, which moves to ∼2.5 km long ice tongue of ideal shape. The ice tongue is angled slightly. Well-expressed LIA moraines stretch along the both sides of the glacier tongue, which makes a curve of ∼1.1 km from the end of the ice tongue and moves to terminal moraine. By these forms of the relief (moraines) it is easy to reconstruct the boundaries of the glacier. The ratio of the lateral moraine crest height to the valley bed height allows us to calculate the glacier thickness. Also there are well-preserved microstade moraines in front of the glacier, which coincides by a minimum error

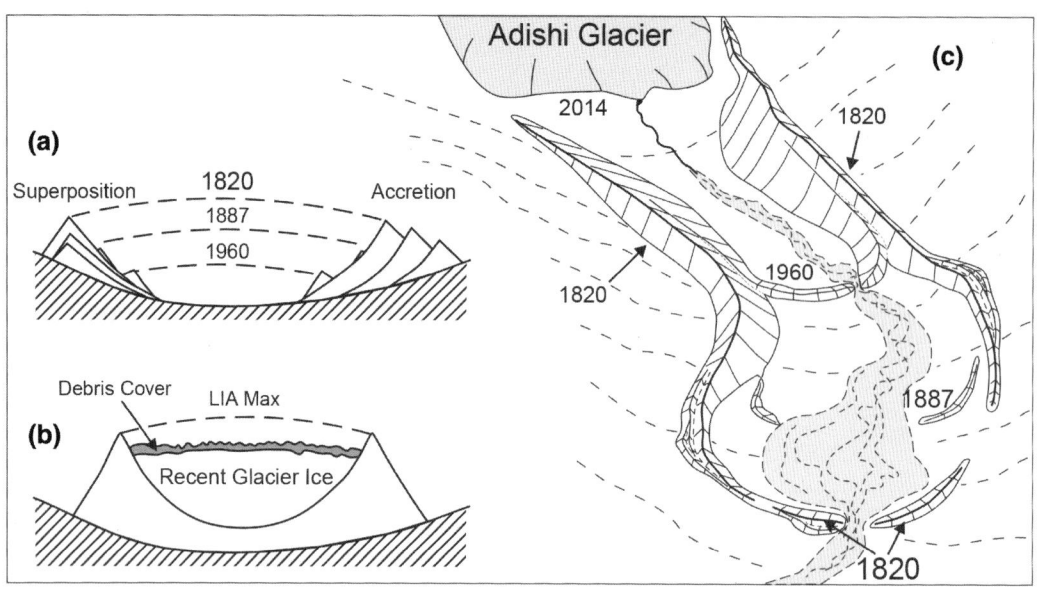

Fig. 5.10 a—Scheme of Adishi glacier LIA moraine depositional structure; two modes of moraine building—by accretion and super position. **b**—Scheme of recent debris-covered glacier; note strong lateral moraine and moraine accretion on the sole of glacier which raises the glacier above valley bottom. **c**—Scheme of typical moraine formation in the forefield of a Adishi glacier, on valley bottom not affected by rough topography

to the glacier contours shaped by us based on topographic maps of 1887–1960. After GPS surveying we made a glacier's melting model from the LIA Maximum until 2014, where there is a glacier reduction both in length and thickness (Fig. 5.10a, b, c).

It has been concluded by obtained result, that during the LIA Maximum the Adishi glacier tongue was of ∼110–120 m width, by the end of the nineteenth century—∼90–100 m width, by the 1960s of the twentieth century—∼70–80 m width (Fig. 5.10a) and in 2014—∼50–60 m. The glacier tongue is covered by the weathered materials of ∼0.1–0.2 m thick (Fig. 5.10b).

Moraine of the LIA Maximum of the Adishi glacier is ideally seen in the ASTER aerial image as well. Therefore, for resulting in the minimum error in parallel with the field works we used the mentioned space image, where we depicted the contours of the glaciers of the years of 1887 and 1960 as well (according to the topographical maps) (Fig. 5.11a, b).

As the survey showed, during the LIA Maximum the Adishi glacier tongue descended to ∼2280 m above sea level. For about next 70 years (1887) the glacier tongue retreated by ∼230 m, while hypsometrically raised by ∼20 m (∼2300). In 1887–1960 the glacier tongue retreated by ∼380 m, while it raised hypsometrically by ∼30 m (∼2330). The fastest raising of the glacier tongue hypsometrically was observed in the years of 1960–2014, when its ice tongue was raised by ∼155 m above sea level, and it was shortened in length by ∼500 m. In total, in the years of 1820–2014 the Adishi glacier area was reduced from ∼14.0 to 9.5 km², while its tongue retreated by ∼1110 m.

In total, we conducted the same works in 24 glaciers on the southern slopes of the Caucasus (Table 5.12). As a result of the study it was found that the highest rates of the retreat is fixed in the compound-valley glaciers (Kvishi, Tviberi, Llekhziri, etc.), the areas of which (of each of them) exceeded 40.0–50.0 km² during the LIA Maximum at, while their ice tongues descended by ∼400–500 m hypsometrical threshold compared to the present one; accordingly, the melting was more rapid. After their separation (1900–1960 years) the rate of their retreat is slowed down.

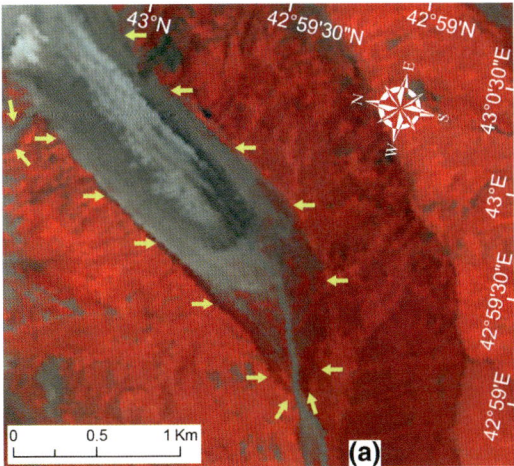

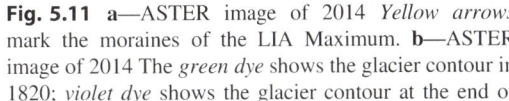

Fig. 5.11 a—ASTER image of 2014 *Yellow arrows* mark the moraines of the LIA Maximum. **b**—ASTER image of 2014 The *green dye* shows the glacier contour in 1820; *violet dye* shows the glacier contour at the end of the nineteenth century (in 1887, according to the topographical map); *yellow dye* shows the glacier contour in 1960 (according to the topographical map); *blue dye* shows the glacier contour in 2014

By small indices are distinguished the simple-valley glaciers (excluding the Khalde and Shkhara glaciers), they did not exceed the 14.0 km^2 during the LIA Maximum. Of special note is a compound-valley type of Khalde glacier (the area—8.8 km^2, the length—7.5 km, height of the ice tongue—2545 m above sea level) located in the Enguri River headwaters (the Khaldechala River). Its retreat after the LIA Maximum is only ~0.8 km. Such a small figure on the Southern slope of the Caucasus is not observed in any other simple-valley or compound-valley glaciers. One of the reasons might be the fact that the Khalde ice tongue is covered by thick weathered materials (~1.0–1.5 m), which protects the ice tongue from the intense melting.

On the basis of the conducted researches we can conclude that the area of the valley glaciers located on the southern slope of the Caucasus reduced by ~25–35% in 1820–2014, while their tongues retreated by ~2.1 km in average. Hypsometrical raising of the ice tongues after the LIA Maximum is characterized by the different parameters. Here a major role plays the morphological conditions and morphometric data of the bottom of the underglacial valley.

The highest hypsometric raise had the glaciers with the high ledges in front of the ice tongues (Tsaneri, Ushba, Dolra, Suatisi, Kirtisho, Didi Edena, and others) and early the ice tongues were hanging over these ledges. Of special note is the compound-valley type of Chalaati glacier (area—8.59 km^2), located in the Enguri River basin, the tongue of which descends to the 1960 m above sea level and intrudes into the forest zone. None of the other glaciers on the southern slope of the Caucasus descends to such a low hypsometrical altitude. The glacier tongue raised by ~340 m hypsometrically after the LIA Maximum.

5.6 Studying the Glaciers Dynamics by the Micro-stade Moraines

In the recent years, much of the world attention has been paid to the study of the glacier dynamics according to the micro-stade moraines. Against the background of the general retreating of the glaciers, a short forward motion, or stationary state of glaciers is observed almost everywhere. The investigations showed that micro-stade moraines are clearly expressed in the glaciers of Georgia, which allows us to make a reconstruction

Table 5.12 Reduction of the valley glaciers in the southern slope of the Caucasus in 1820–2014

| The first group of glaciers | | | The second group of glaciers | | | The third group of glaciers | | |
Name	Morphological type	Retreat (~km)	Name	Morphological type	Retreat (~km)	Name	Morphological type	Retreat (~km)
Kvishi	Compound-valley	5.5	Ushba	Compound-valley	1.9	Banguriani	Cirque-valley	1.4
Tviberi	Compound-valley	4.4	Sopruju	Cirque-valley	1.8	Tsitela	Hanging-valley	1.4
Lekhziri	Compound-valley	3.8	Sakeni	Simple-valley	1.8	Psishi	Simple-valley	1.3
Dolra	Compound-valley	3.3	Kirtisho	Simple-valley	1.7	Marukhi	Simple-valley	1.3
Eastern Suatisi	Compound-valley	3.3	Shdavleri	Simple-valley	1.7	Shkhara	Compound-valley	1.2
Tsaneri	Compound-valley	3.0	Tbilisa	Simple-valley	1.6	Adishi	Simple-valley	1.1
Buba	Compound-valley	2.7	Boko	Simple-valley	1.6	Guli	Simple-valley	1.0
Chalaati	Compound-valley	2.1	Chanchakhi	Simple-valley	1.5	Khalde	Compound-valley	0.8

of variation of the glaciers in time and space, identify the number of micro-stages and limits of the glaciers expansion. The micro-stade moraines allow assuming the intra-secular variability of climate (Gobejishvili 1995).

In order to obtain the quantitative parameters to describe the dynamics of modern glaciers, together with the field materials and modern aerial images, we used the data of special photo-theodolite surveys of many glaciers. We decoded the aerial images and through them identified the number and spatial state of the micro-stade moraines. In addition, by using the field work, literary and cartographic materials, we identified the time of origin of micro-stade moraines. Below we give a detailed description of the dynamics of some of the glaciers.

Zopkhito-Laboda glacier. The glacier has well-expressed lateral moraines, which are transformed into the LIA maximum terminal moraine ∼ 1500 m away from the glacier tongue. Seven micro-stade moraines are well fixed in the valley, which, against the background of the glacier retreat, corresponds to the separate stages of the micro-formations or stationary state. The width of the micro-stade moraines is ∼ 1–3 m. They are arch-like hillocks, which in some cases are leaned against a stade moraine. Sometimes, their trace is wiped off the relief, or is violated due to the exogenous processes. This is why the identification of the shape of the glacier tongue or calculation of its area or volume during the individual micro-stages is often complicated. We identified the distance from the glacier tongue to the micro-stade moraines and among the micro-stade moraines, absolute heights of the moraines and relative heights in relation to each other.

The work of Dinik (1890) is very interesting for studying the dynamics of the Zopkhito glacier, which describes the state of the glacier in 1882 (Gobejishvili 1995). The configuration of the glacier described by him corresponds to the shape of the micro-stade moraine, and the glacier tongue itself is located above it. A flow, coming out of Brili glacier, could get on the glacier surface only during its being in such a state.

By our data, at this time the II micro-stade moraine was formed and the Zopkhito glacier

was retreating. Therefore, the II micro-stade moraine was formed in ∼ 1870–1880. In the same years the II micro-stage moraine was formed in the Bezing glacier—the largest glacier of the northern slope of the Caucasus. It was in this period when the solar activity was reduced as well (Gobejishvili 1995).

The analysis of the materials of the topographic survey conducted in 1889–1992 allows considering the years of ∼ 1890–1895 as a time of the III micro-stade moraine formation. The water flowing out of the Brili glacier does not fall onto the surface of the Zopkhito glacier anymore and joins directly to the Zopkhitura River near the glacier tongue. At that time the glacier tongue was ended at the height of ∼ 2180 m. This height well corresponds to the height of the III micro-stade moraine (∼ 2187 m).

To identify the time of the V micro-stade moraine formation (with the height of ∼ 2270 m), the data of Tsereteli (1959) are of a particular importance. In 1937, in front of the glacier tongue, he made a mark on the moraine hillock with the height of ∼ 2270 m, which was ∼ 120 m away from the glacier; this hillock well corresponds to the location of the V micro-stade moraine.

The observations on the glaciers of Caucasus in 1930–1932 showed that at that time they were in the stationary state, or were characterized by the temporal advancements. In this period, the V micro-stade moraine was formed and the solar activity was reduced as well. Tian Shan glaciers advanced in the same period (Gobejishvili 1995). All the above-listed facts allow considering the years of 1930–1935 as the time of the V micro-stade moraine formation.

By the data of the meteorological devices (Mamisoni, Oni, Shovi, Ghebi), the years of 1939–1941 were marked with abundant of atmospheric precipitations, particularly in the cold period, when the amount of precipitations exceeded the annual rates by 200 mm/year. Tsereteli (1959) considered that in 1945–1952, the rate of glaciers retreat was reduced for several times as compared to the rates in the previous years. It should be noted that in 1946–1950 the glaciers advancement was observed both in the Caucasus and Tian Shan as well (Gobejishvili 1995). Based on these materials, we consider that the VI micro-stade moraine was originated in 1945–1952. The accuracy of this date was proved later by the investigations carried out in the Svaneti glaciers by Golodovskaya (1984) by the lichenometric method.

We do not have the direct data about the IV micro-stage moraine. In 1907–1908 and 1911–1913 the glaciers of the Caucasus were in a stationary state, while in the same period the Tian Shan glaciers were advancing (Gobejishvili 1995). If we take into account that the formation of the micro-stade moraines (II–III, V–VI) of the Zopkhito glacier is subject to a ∼20–25-year-long cycle, then the IV micro-stade moraine would have been originated in ∼1910–1917.

By using the micro-stade moraines, topographic maps of the 19–20th centuries and Landsat aerial images, it will be interesting to consider the dynamics of the Zopkhito-Laboda glacier in the years of ∼1820–2014, ∼1820–1890, ∼1890–1960 and ∼1960–2014 (Fig. 5.12).

The Figure shows that the retreat of Zopkhito much exceeds the same data of Laboda, particularly in the last 54-year-period, when in case of Zopkhito the indices are 3 times higher. Here we deal with the glacier morphology and orographic location, namely: the Zopkhito glacier, as we mentioned above, is of a south-eastern exposition and its tongue is slightly inclined; it is bordered by a rocky relief from the north, west and south and has no relief barrier from the east. Therefore, the sunshine duration is quite high here, so as the falling angle of the solar beams than in the Laboda glacier, which is of a south-western exposition and is bordered by high rocky relief from the east, north and west. Therefore, the sunshine duration is much less here. Besides, the ice tongue of Laboda is inclined twice as much, as compared to Zopkhito; respectively, the falling angle of the soar beams is lower and the melting course is less intense.

Besides, we can single out one reason because of which the Zopkhito glacier melting exceeds

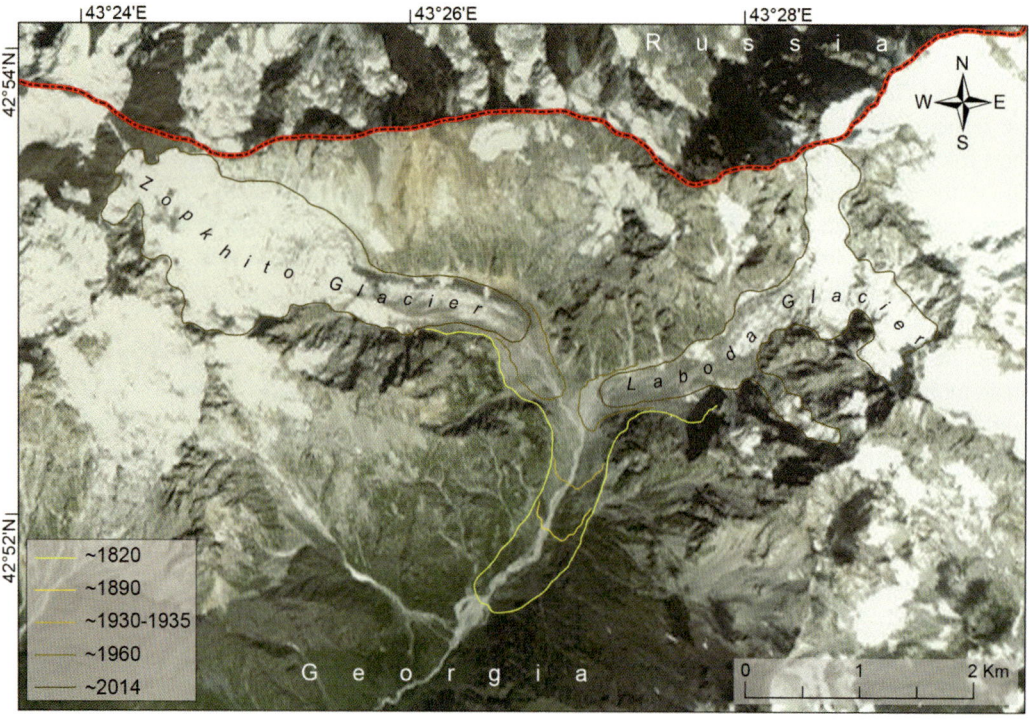

Fig. 5.12 Dynamics of the Zopkhito-Laboda glacier in ∼1820–2014

Table 5.13 Dynamics of the Buba and Adishi glaciers according to the micro-stade moraines

Micro-stages	Moraine (~age)	Glacier among the micro-stages			
		Distance (~m)	Area (~km^2)	Mean thickness (~m)	Capacity (~m^3)
1. Buba	1984				
VII	1970–1973	130.0	0.09	8.0	0.006
VI	1945–1950	255.0	0.11	10.0	0.008
V	1930–1935	415.0	0.22	20.0	0.016
IV	1910–1935	120.0	0.07	6.0	0.005
III	1890–1895	300.0	0.19	15.0	0.012
II	1870–1875	240.0	0.11	10.0	0.008
I	1845–1850	150.0	0.07	6.0	0.005
LIA maximum	1820	70.0	0.03	3.0	0.002
2. Adishi	1986				
VII	1970–1975	60.0	0.03	5.0	0.004
VI	1945–1950	75.0	0.06	10.0	0.008
V	1930–1935	135.0	0.12	15.0	0.012
IV	1910–1935	55.0	0.10	5.0	0.004
III	1890–1895	115.0	0.20	20.0	0.016
II	1870–1875	70.0	0.14	10.0	0.008
I	1845–1850	90.0	0.10	10.0	0.008
LIA maximum	1820	65.0	0.02	5.0	0.004

the melting of the Laboda glacier. In the last decades the area of the territory covered with weathered material is increased in the surface of the Zopkhito glacier tongue and as it is not characterized by a sharp inclination; the material brought from the lateral slopes are widely spread on the glacier tongue surface thus increasing the melting intensity. The same cannot be said about the Laboda glacier, because due to the great inclination of the glacier tongue, the glacier surface is clear and has no weathered material on it; consequently, the solar beam reflection is high and the melting rate is slowed down.

R. Gobejishvili studied the micro-stade moraines of the following glaciers: Southern Marukhi, Klichi, Dolra, Boko, Buba, Tbilisa and etc. The conducted investigations showed that 6–7 micro-stade moraines are well expressed in the basins of all glaciers (Gobejishvili 1981, 1995). The equal number of the micro-stade moraines indicates that against the background of the general retreat of the glaciers, short-term advancements were observed, or they were in the stationary state, during which the moraine hillocks were originated. In the micro-advancement of the glaciers can be observed the synchronicity and it is repeated in every ~20–25 years.

Golodovskaya (1984) studied the glaciers of the Caucasus in the 1980s by a lichenometric method. The comparison of the data by R. Gobejishvili to the results obtained by this method once again assured us that the fluctuation of the Caucasus glaciers was increased synchronously. This is why the age of the micro-stade moraines is either coincided or very close to each other.

The conducted studies showed that after the LIA maximum the glaciers of Georgia retreated unequally despite the complete synchrony in the location of the micro-stade moraines (Table 5.13). The distance between the IV–V micro-stade moraines of all studied glaciers is less than the distance between the III–IV and V–VI micro-stade moraines. This once again indicates that the formation of the micro-stade moraines in the glaciers of Georgia occurred at the same time.

The analysis of the distribution of the micro-stade moraines allows identifying not only the longitudinal reduction of the glaciers, but also their reduction in volume and area.

References

Catalog of Glaciers of the USSR (1975) Katalog Lednitov USSR, vol 8–9. Gidrometeoizdat, Leningrad, 108 pp

Dinnik NY (1890) Present and old glaciers of Georgia. KOIRGO 14:88–102, Tiflis (in Russian)

Dyurgerov MB (2005) Glacier mass balance and regime data of measurements and analysis. Institute of Arctic Alpine Research, University of Colorado, Boulder, Occasional Paper No. 55, 268 pp

Furrer G, Burga C, Gamper M, Holzhauser H, Maisch M (1987) Zur Gletscer, Vegetations und Klimageschichte der Schweiz seit der Spateiszeit. Geogr Helv 42 (2):61–91

Gobejishvili RG (1981) A study of modern relief-forming processes in mountain areas by stereophotogrammetric methods (on the example of Racha in western Georgia). Publishing House "Metsniereba", Tbilisi (in Russian)

Gobejishvili RG (1989) Glaciers of Georgia. Publishing House "Metsniereba", Tbilisi (in Russian)

Gobejishvili RG (1995) The evolution of the modern ice age glaciers and mountains of Eurasia in the Late Pleistocene and Holocene. The thesis of doctor of science degree in geography (in Georgian)

Golodovskaya NA (1984) lixenometria i kolebania lednikov za poslednie 700 let. v kh. kolebanie lednikov k prochessi morenonakoplenia na centralnom kavkaze (Lichenometry and glacier fluctuations over the past 700 years. In: Glacier fluctuation). M. Nauka (in Russian)

Grove JM (1988) The little ice age. Metchuen and Co. Ltd., London. ISBN 0-415-01449-2

Hock R (1999) A distributed temperature-index ice- and snowmelt model including potential direct solar radiation. J Glaciol 45(149):101–111

Holzhauser H (1983) Die Geschichte des Grossen Aletschgletschers Wahrend der Letzten 2500 Jahre. Bull Murithienne 101:113–134

Hunt BG (2006) The medieval warm period, the little ice age and simulated climatic variability. Clim Dyn 27:677–694

Ivankov PA (1959) Oledinenie bolshogo Kavkazai ego dinamika za godi 1890–1946 (The glaciation of the Greater Caucasus and its dynamics during the years 1890–1946). Izv. VGO. T. 91, Vip. 1 (in Russian)

Khazaradze RD (1971) Relief, kontinentalnie otlojenia i pleictocenovie oledenenie bac. R, Inguri (Relief, continental sediments and Pleistocene glaciation Enguri River basin), Avtoreferat kand. Disertacii, Tbilisi (in Russian)

Khromova T, Nosenko G, Kutuzov S, Muraviev A, Chernova L (2014) Glacier area changes in Northern Eurasia. Environ Res Lett 9:015003. doi:10.1088/1748-9326/9/1/015003

Kotlyakov VM, Krenke AN (1981) Glaciation actualle et climat du Caucase. Rev Geogr Alp 69:241–264

Kovalev PV (1961) Sovremennie i drevnie oledenenie bacceina r. Inguri (Modern and ancient glaciation Enguri River basin). Mat. Kavkaz. Eksped. T. 2. Xarkov. Izd XGU (in Russian)

Podozerskiy KI (1911) Ledniki Kavkazskogo Khrebta (Glaciers of the Caucasus Range): Zapiski Kavkazskogo otdela Russkogo Geograficheskogo Obshchestva. Publ. Zap. KORGO., Tiflis, 29, 1, 200 pp (in Russian)

Shahgedanova M, Nosenko G, Kutuzov S, Rototaeva O, Khromova T (2014) Deglaciation of the Caucasus Mountains, Russia/Georgia, in the 21st century observed with ASTER satellite imagery and aerial photography. The Cryosphere 8:2367–2379. doi:10.5194/tc-8-2367-2014

Solomina ON (2000) Retreat of mountain glaciers of Northen Eurasia since the Little Ice Age maximum. Ann Glaciol 31:26–30

Tielidze LG (2014) Glaciers of Georgia. Monograph. Publ. House "Color", Tbilisi, 254 pp (in Georgian)

Tielidze LG (2016) Glacier change over the last century, Caucasus Mountains, Georgia, observed from old topographical maps, Landsat and ASTER satellite imagery. The Cryosphere 10:713–725. doi:10.5194/tc-10-713-2016

Tsereteli DV (1959) Izmenenie lednikov iuzhnovo sklona Kavakiona za poslednie 25 let (Changing the southern slope of the Caucasus glaciers over the past 25 years). Coobsh. An. GSSR. T. 21. #6. Tbilisi (in Russian)

Tsomaia VS (1960) k voprosy o dvijenii lednikov kavkaza (to the question on the motion of the Caucasus glaciers). Tr. Tbil. NIGMI, vip. 7 (in Russian)

Tsomaia VS, Drobyshev OA (1970) The results of glaciological observations on the glaciers of the Caucasus. "ZakNIGMI", no. 45 (in Russian)

Vartanov GS (1978) Issledovanie gornix lednikov kartograficheskim, geodezicheskimi i stereofotogrammetricheskimi metodami, na primere basseina r. Inguri (Study of mountain glaciers cartographic, geodetic, and stereo photogrammetric methods on the example of river basin Enguri), Avtoreferat kand. Disertacii, Tbilisi (in Russian)

Zasadni J (2007) The Little Ice Age in the Alps: its record in glacial deposits and rock glacier formation. Studia Geomorphologica Carpatho-Balcanica Xll:117–137. http://paleobiology.si.edu/geotime/main/htmlversion/pleistocene3.html

Part II

Evolution of Glaciations in the Late Pleistocene and Holocene

Late Pleistocene and Holocene Glaciation

6

Abstract

The Part 2 includes the information on Late Pleistocene and Holocene glaciation of Georgia. Reconstruction of old glaciation is conducted using the analogy approach both in the Caucasus main range and in the southern highland of Georgia.

Keywords

Late Pleistocene · Holocene · Glacier reconstruction

The Pleistocene geological record gives evidence of 20 cycles of advancing and retreating continental glaciers, though during most of the Pleistocene glaciers were far more extensive than they are today. Much of this glaciation occurred at high latitudes and high altitudes, especially in the Northern Hemisphere. Up to 30% of the Earth's surface was glaciated periodically during the Pleistocene. Large portions of Europe, North America (including Greenland), South America, all of Antarctica, and small sections of Asia were entirely covered by ice. In North America during the peak of the Wisconsin glaciation approximately 18,000 years ago, there were two massives yet independent ice sheets. Both the eastern Laurentide and the western Cordilleran ice sheets were over 3900 m thick. In Europe, the ice covered Scandinavia, extended south and east across Germany and western Russia, and southwest to the British Isles. Another ice sheet covered most of Siberia. In South America, Patagonia, and the southern Andes mountains were beneath part of the Antarctic ice sheet (paleobiology.si.edu). One of the important mountain ranges in the Eurasia was the Caucasus, which was covered with quite thick glacier cover during the Pleistocene.

After the famous scheme—the Paleogacial Concept by Penck and Bruckner[1] (1901/1909) almost all researchers tried to compare the Caucasus to the Alps, where the old glaciers were emerged in the plain from the foothill zone. Drawing such a picture for the Caucasus was acceptable for the researchers. They did not took into account that the Alps and the Caucasus are located at different latitudes, and what was common to the Alps at a spatial scale, could not be identical to the Caucasus. Surely, the glaciation both, in the Alps and Caucasus was of coincided character. Coincided was the fact as well, that the glaciation of that period was the strongest, but their expansion limits could not be identical. There were many opponents to this view, who considered that comparing the Alps with other mountainous regions was inappropriate.

Solving this issue is problematic and urgent even today.

6.1 Evolution of the Glaciers in Georgia

One of the problematic issues of the studies of modern geographical science is the study of the paleoglacial issues. The changes in the natural physical-geographical conditions in the earth surface in the Upper Quaternary and Holocene are much depended on the glaciation dynamics. The glaciers in this period changed not only the structure of the geosphere, but they also established new relationship between the geographical elements. It is known that in the Quaternary Age (Table 6.1) the earth surface was subject to multiple glaciations and there are many views

about the limits of their extension. Sometimes these views complement each other, but sometimes they are mutually exclusive.

Basically, all research methods related to the reconstruction of the old glaciation are based on the analysis of the morphosculptural forms created by glaciers. The analysis of the glaciogenic forms does not always give the desirable result, particularly, when dealing with multiple glaciations. Moraines, as the forms of multiple glaciations, are not always maintained in the relief, or sometimes, a very strong glaciation removes the trace of the previous less strong glaciation from the earth surface. There are many such examples with the Caucasus and other mountainous massifs, where the most recent Wurm glaciation removed the real traces of Mindel[2] and Riss[3] glaciations.

At the present stage the glacial and fluvio-glacial forms and deposits in the region of the Caucasus Mountain Range are quite common indicating the quite large scales of glaciers extension. Particularly well preserved are the glacial forms in the basins of the major rivers of the Central and Western Caucasus and on the slopes of the Caucasus branch ranges. Good distribution of the terminal moraines, existence of the associated trough forms and sometimes the allocation of the lateral moraines on the slopes of the river gorges, as well as the well preserved cirque forms allow identifying the borders of the glaciers extension quite accurately.

The researchers use various methods to identify the borders of old glaciations, such as: geomorphological, glaciological, petrographical, lithodynamical, dust analysis, paleobiological, absolute age method, etc. All of these methods allow studying the glaciation according to the

---

[1] In the Alps, Penck and Brückner (1901/1909) concluded that all the glacial landforms they had found could be traced back to four major Pleistocene glaciations (Günz, Mindel, Riss, Würm). This view was generally accepted, and Penck and Brückner's stratigraphical scheme was applied on a global scale. Only very gradually it became clear, that the Alpine stratigraphy was not the key to all glacial sequences, and that glaciation had set in much earlier than the pioneers of Quaternary research had suspected. More and ever older glacial deposits were identified in Iceland, South America, Antarctica and Greenland (Ehlers and Gibbard 2008).

[2] The Mindel glaciation is the third oldest glacial stage in the Alps. Its name was coined by Albrecht Penck and Eduard Brückner, who named it after the Swabian river, the Mindel. The Mindel glacial occurred in the Middle Pleistocene.

[3] The Riss glaciation is the second youngest glaciation of the Pleistocene epoch in the traditional, quadripartite glacial classification of the Alps. The literature variously dates it to between about 300,000 to 130,000 years ago and 347,000 to 128,000 years ago. It coincides with the Saale glaciation of North Germany.

Table 6.1 Subdivisions of the Quaternary period (Encyclopedia britannica 2013)

System	Series	Stage	Age (million)
Quaternary	**Holocene**		0–0.0117
	Pleistocene	Tarantian (Late)	0.0117–0.126
		Ionian (Middle)	0.126–0.781
		Calabrian (Lower)	0.781–1.806
		Gelasian (Lower)	1.806–2.588
Neogene	Pliocene	Piacenzian	Older

different river basins. All of these methods have certain gaps and may lead a researcher to rough mistake.

The material gained through these methods has been interpreted on a single methodological basis known in the literature as Alpine Scheme.

As far back as in 1956 L. Maruashvili, on assessing critically the study of old glaciers in the Caucasus in his monograph (Maruashvili 1956), indicates the mistakes accompanied their works. L. Maruashvili criticizes the Alpine school followers—Reinhardt (1937) and Vardaniants (1937), who, based on the Alpine Scheme, compiled the paleoglacial map of the Caucasus (a paleographic scheme of the Caucasus in the glaciation epoch). These materials were the basis for all following researchers. This view was unanimously shared almost by everyone.

The Alpine Scheme was opposed by L. Maruashvili, who refused to adapt and disseminate it to the Caucasus and who stated that the glaciation in the Caucasus was weaker than it was stated by the representatives of this school. L. Maruashvili considered that in the Quaternary the glacier tongues flowed down to the height of ∼ 1000–1500 m in the Western and Central Caucasus (Bzipi–Kodori basin) and to ∼ 2000–2500 m in the Eastern Caucasus (Tergi, Asa). The snow line depression made ∼ 600–800 m.

In the opinion of Tsereteli (1966), the glaciation of the Caucasus can be compared to that of the Alps and even to other mountainous regions. In his opinion, the course of the glaciation in the Caucasus is behind the Alps glaciation by a half cycle. He considered that the glaciers in the Caucasus descended quite below (if judging by Tsebelda moraine), and the snow

line depression made ∼ 1200 m. This opinion is close to all opinions about the Alps and to our mind, it is generally true. As for the descending of the glacier tongues in the basins of the rivers of Enguri and Rioni to ∼ 800–1000–1100 m, this point of view is correct and is proved by the existence of the moraine complex there.

The Nenskra, Mestia, Dolra, and Ushguli moraines are located at the different altitudes in the Enguri River basin. Due to this, D. Tsereteli considered them to be of different ages (stages). To our mind, this opinion needs further specification.

Khazaradze (1971) studies the questions of glaciation using the petropraphic method. His works give the borders of the glacier extension in the basins of the rivers of Georgia with the locations of the erratic boulders.

The old Caucasus glaciation is considered in the works by Abich (1865), Reinhardt (1936, 1937), Kuznetsov (1931), Vardaniants (1930, 1937), Klopotovskiy (1949), Paffenholz (1958), Nemanishvili (1962), Dondua (1959), Astakhov (1973), Milanovskiy (1960), Dumitrashko (1961), Kovalev (1961), Tabidze (1965), Tsereteli (1966), Serebruanny and Orlov (1989), Gobejishvili (1995), etc.

There are two major schools established around the old Caucasus glaciation: (1) the followers of the Alpine school think that the glaciation during the Wurm was of a sheet or semi-sheet with the snow line depression of ∼ 1100–1300 m and the glacier tongues descending to a very low hypsometric level. The glaciers in the river gorges joined each other and emerged in the foothill zone as a single tongue (the rivers of Tergi, Enguri, Baksani, Chereki,

Kubani, Kodori, etc.); (2) By the opinion of another group of scientists, the glaciation was weak during the Wurm, with the snow line depression at ~600–800 m, and the glacier tongues descending to ~1500–2000 m (from Kodori to Tergi) (Maruashvili 1956).

As it was mentioned above, there are two different opinions regarding glaciation in the Caucasus. Both opinions are based on certain factual material and at one glance, it is difficult to assess which opinion is right, though, we think that there is a truth in both opinions and their comparative analysis will yield in the desirable result.

For this purpose, we decided to study these questions using a new approach. For the study we used the geomorphological, cartographical, petrographical, and aerial images decoding methods (in order to identify the areas of the distribution of moraines). The complex application of these methods led us to the different method of study, which will allow identifying the limits of the old glaciation according to the individual orographic units and the river basins, which finally gives the full picture of the Wurm glaciation extension in the Caucasus. Based on these data we compiled the map of the Wurm glaciation of the Caucasus (Gobejishvili and Tielidze 2012).

During developing the above-mentioned method, we used the analog method, well known in scientific literature. The essence of the analog method is as follows: in case of similar natural conditions (factors), the same process or phenomenon can be studied in one region and apply the obtained information to another region. The analog method is widely used in glaciology and geomorphology. For instance, let us consider the glaciers located on the southern slope of the Caucasus. If considering a valley glacier and studying its thawing or accumulation, we will see that the numerical data typical to it can be applied to the similar glaciers found in the adjacent basins. The glaciologists call such a glacier the model glacier.

When reconstructing the morphology and sizes of the old glaciation, an objective difficulty is that the trace of old glaciation is slightly preserved in the relief and many sedimentary forms very much look like the glacial forms by their appearance. The trough form of the valley at many locations is modified due to the weathering, erosion or denudation processes. Under their impact, the lateral moraines on the slopes are absolutely removed, or are survived as fragments. The terminal moraines are preserved not so well, as they are washed off under the action of fluvio-glacial or mudflow currents. Due to the active exogenous processes, the old glacial material is transported to the lower hypsometric levels and forms a false image of the real extension of the glaciers.

Despite the many-year detailed studies, the genesis of these formations is impossible to specify using the traditional glacial-geomorphological approach. Because of this there are essential differences in the identification of the morphology and scale of the glaciation in many mountainous regions of the world.

It is important that the data on modern glaciers are not used in solving the paleoglacial problems. In order to fill this gap, we studied the modern glaciers of Georgia.

We identified the link between the length of the modern glaciers and their feeding firn basins. The analysis of aerial images and cartographic materials showed that the tongues of the valley glaciers are fed from the firn within the vast cirque forms located at higher altitudes. The larger the firn basin is, the longer the glacier is. If the glacier of a certain length corresponds to the feeding area of the modern glacier, it is natural that the feeding area of the old glacier would have formed a glacier of a certain length (Gobejishvili et al. 2011). This idea can be expressed by a directly proportional dependence formula:

$$\frac{Lc}{Sc} = \frac{Ld}{Sd}$$

where the *Lc*—is the length of the modern glacier, *Sc*—is the area of the firn basin of the modern glacier, *Ld*—is the length of the Pleistocene glacier and *Sd*—is the area of the firn of the Pleistocene glacier.

In this formula, the values of *Lc* and *Sc* are identified using the topographic maps (Gobe-jishvili et al. 2011) for the valley glaciers of some mountains in Georgia and Eurasia (Tian Shan, Spitsbergen, Tibet, Himalayas).

For comparison purposes, we classified the gained quantitative indexes by their values and grouped them. We singled out four groups of glaciers, which clearly differ from each other with the above-mentioned dependence. We denoted this dependence by coefficient *K*:

$$\frac{Lc}{Sc} = K$$

The first group included the hanging-valley and simple-valley glaciers; for them *K* = 0.81 (with the correlation coefficient of 0.96); the second group included the simple-valley glaciers with a multi-chamber firn with *K* = 0.50 (0.95); the third group included the compound-valley glaciers with *K* = 0.33 (0.99); and the fourth group included the branched glaciers with *K* = 0.13 (0.86).

Today there are no glaciers of the fourth group found in the Caucasus. Classical examples of such glaciers are Inilchek, Koendi, and Mushke-tov in Tian Shan, Rongbuk—in the Himalayas, etc. Further division of the glaciers of this group for the other mountainous regions (Alaska, Karakorum, Hindu Kush) will be necessary to improve the accuracy of the data (according to the number of branches composed the glacier). For the higher accuracy, the coefficient is better to calculate for the individual types of the glaciers.

Based on the data gained for individual groups, a nomogram showing the relation between the area of the glacier firn basin and the glacier length was drafted. After defining coefficient (*K*) the proportional dependence formula will be as follows:

$$\frac{Ld}{Sd} = K, \quad then \, Ld = Sd \times K.$$

In this formula, the *Sd*—is the feeding area of the old glacier, i.e., it is the cirque of the old glacier, which was the source of feeding the glacier.

The morphological and morphometric analysis of the relief and processing of modern aerial images showed that the old cirques developed in the axial zone of the Caucasus built with crystal rocks are well preserved in the relief. The old cirques are mostly originated during the glaciation of the Late Pleistocene. This glaciation is known as the Wurm Glaciation in literature. Post-Wurm stage glaciation led to the slight redeepening of the existing cirque. We mapped almost all glacial cirques and measured their areas. In addition, we identified the height of the lower limits of the cirques. To our opinion, the height of the lower limits of the cirques is the indicator of the firn line location. Based on the gained materials, we calculated the lengths of the Late Pleistocene glaciers, the altitudes of their tongues, and location of the firn lines.

In order to be sure in the reliability of the mentioned method, we examined it in the Enguri River gorge, where the trace of the old glaciation is well preserved in the relief. Choosing the Enguri River gorge as a model object was identified by the following factors:

1. The traces of the Wurm (Late Pleistocene) and Post-Wurm glaciations are well preserved in the Enguri River gorge and its tributaries in the form of glacial cirques, moraines, and troughs;
2. There are all morphological types of the modern glaciers typical to the Caucasus in the Enguri River basin;
3. Other researchers have also reconstructed the old glaciation in the Enguri River basin (Kovalev 1961; Tsereteli 1966; Khazaradze 1971; etc.).

We conducted the detailed glacial-geomorphological studies, decoded the aerial images, and mapped the old glacial forms. All the above-listed is the reliable basis to check the data obtained by our method.

We mapped the old glacial morphosculptural forms (troughs, cirques, moraines, river terraces) in the Enguri River basin with the scale of 1:50,000 and 1:200,000 by decoding the modern aerial images and compiled their distribution map (Fig. 6.1).

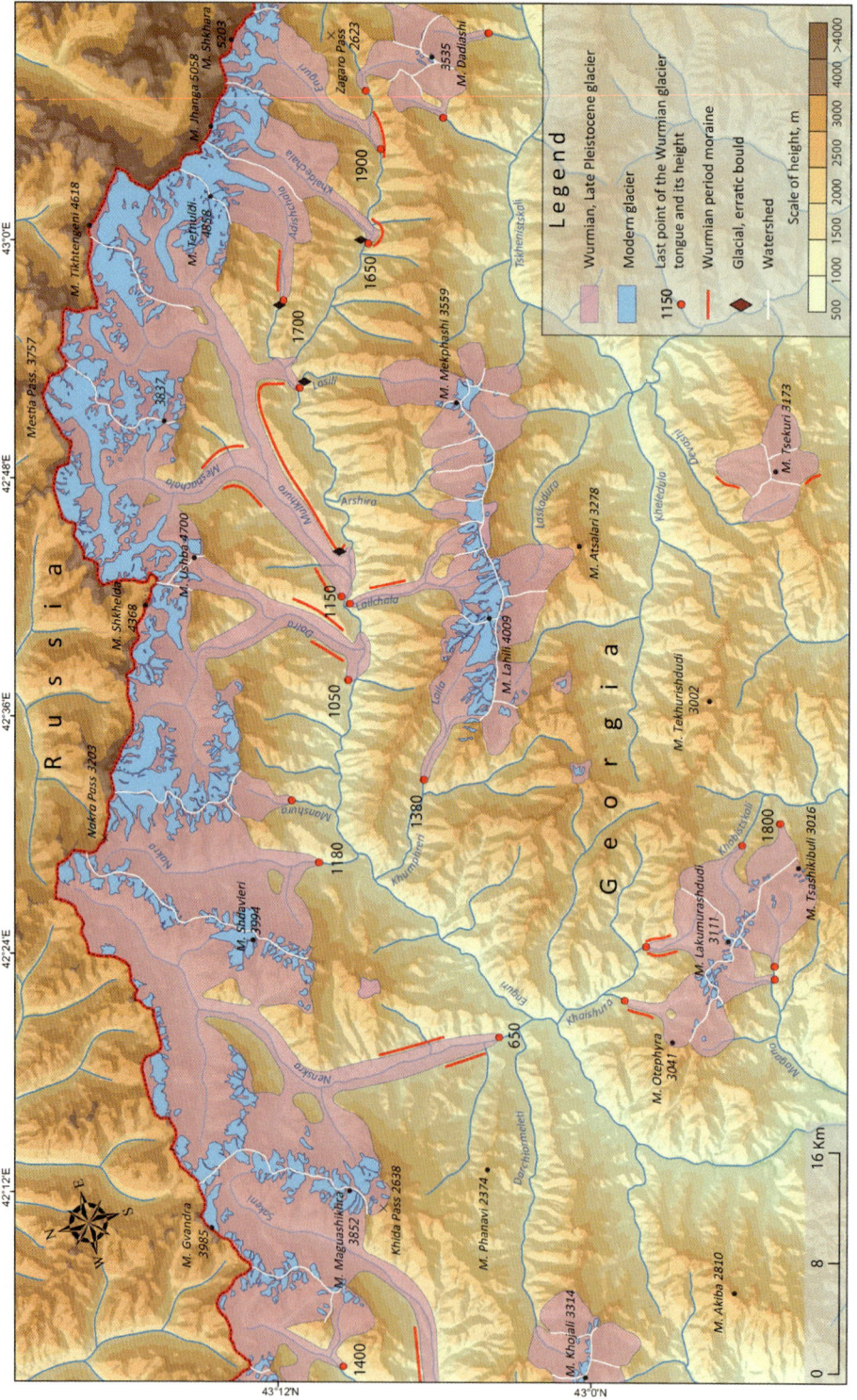

Fig. 6.1 Modern and Late Pleistocene (Wurmian) glaciers in the Enguri River basin

The conducted studies showed that the old cirques are very well seen not only in the watershed of the Caucasus, but also in its branch ranges.

When mapping the glacial forms in the Enguri River gorge using the topographic maps and aerial images, we used the ArcGIS software to identify the heights of the lower thresholds of the old cirques. As we already mentioned, their heights fix the location of the firn line in the Late Pleistocene.

The lower threshold of the old cirques in the Enguri River basin is located in the lowest altitude (~ 2000 m) in the Samegrelo range. It is located at the altitude of ~ 2100 m in the Nenskra River basin and on the northern slope of the Svaneti Range. The lower thresholds of the old glaciers are at the altitude of ~ 2200 m in the basins (in the headwaters) of the rivers of Nakra, Dolra, Mulkhura, and Enguri (Table 6.2).

By considering the morphological types of the glaciers in the Enguri River basin, we have identified the four basins: the Mulkhura River basin, presented by compound-valley glaciers; the Dolra River basin with spreading of the compound and simple-valley glaciers; the Adishchala River basin with the valley glaciers; and the Nenskra River basin with a great number of valley and cirque glaciers.

There were two old cirques identified in the **Mulkhura River basin**, one in the headwaters of the Mestiachala River and another in the headwaters of the Mulkhura River itself. However, the morphological analysis of the relief showed that there are two independent cirques of Chalaati and Lekhziri in the Mestiachala, and there are two independent cirques of Tviberi and Tsaneri in the headwaters of the Mulkhura itself. After combining all of these glaciers, we had a branched glacier composed of four flows in the form of a single ice-tongue in the Mulkhura gorge. The length of the glacier was ~ 31.5 km (Fig. 6.2); according to the glacial-geomorphological observation materials, the greatest length of the glacier is ~ 38.0 km. Difference among them is 6.5 km (8.2%). The length of the glacier in the Mestiachala gorge (Lekhziri paleo-glacier) was ~ 33.7 km using the above-mentioned method and it was ~ 32.5 km using the glacial-geomorphological method making the difference of 1.2 km (3.7%). The glacier was of a compound-valley type; the glacier tongue in Latali surroundings ended at the altitude of ~ 1150 m. Indeed, a lateral moraine is well represented in the surroundings of the village of Latali, and the valley retains a trough-like shape up to this point (Gobejishvili et al. 2012).

Table 6.2 Morphological indicators of the Late Pleistocene glaciers in the Enguri River basin

Glacier name	Old glacier cirque area, $\sim$ km^2	Calculated glacier length, $\sim$ km	Glacier length by terminal moraine, $\sim$ km	Height of the lower threshold of cirque (firn line) above sea level, $\sim$ m	Height of the end part of the ice-tongue above sea level, $\sim$ m
Nenskra	275.0	36.0	40.0	2100	650.0
Dolra	105.0	34.5	33.0	2200	1050.0
(a) Mulkhura, general	270.0	35.1	38.0	2200	1150.0
(b) Lekhziri	102.0	33.7	32.5	2200	1150.0
Adishi	24.0	19.4	18.5	2200	1700.0
Khalde	31.0	15.5	16.0	2200	1650.0
Enguri (Shkhara)	35.0	17.5	17.0	2200	1900.0
Lailchala	23.0	11.5	12.0	2100	1100.0
Laila	17.0	13.7	13.0	2100	1380.0

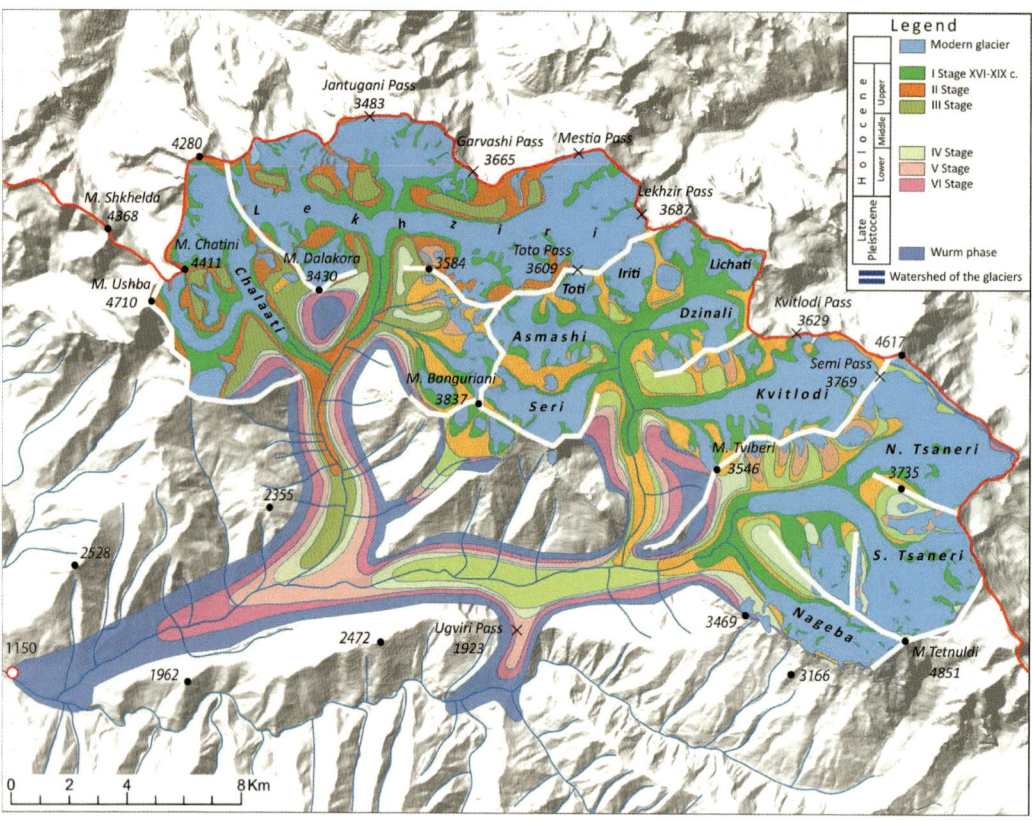

Fig. 6.2 The Late Pleistocene and modern glaciers in the Mulkhura and Mestiachala River gorges

The traces of the terminal moraines can be found on the left side of the Enguri River, in the territory of the village of Laila. This indicates that the glacier tongue during the Wurm had a fan-like shape at the end. The signs of the terminal moraine can be found in the surroundings of the village of Lakhushti within ~0.5–1.0 km along the Enguri River. A well expressed in the Zagari range lateral moraine, which ends at the vicinities of the village of Soli and does not extend to Latali with the morphological characteristics of moraine, though there is a moraine material on the slope, is a good sign of the fact that the Tsaneri glacier could not reach Latali. Consequently, we can conclude that the lengths of the glaciers calculated using the mentioned method are reliable.

Thereafter, it was necessary to us to examine this method for the basins presented by a single valley type of glacier. Such river basins are

Adishchala, Khaldechala and headwaters of the Enguri River itself. As the river Adishchala gorge is studied geomorphologically most thoroughly, we considered that the examination of the method with this river was most advisable. A whole range of Wurm and Post-Wurm glaciation is presented in the **Adishchala gorge** in the form of the moraines (lateral moraines, fragments of the terminal moraine).

During the Wurm, the length of the Adishi glacier was ~19.4 km. The glacier of this length descended ~4 km below the village of Adishi and ended at a big stone. Morphologically, there are signs of the terminal moraines of the Wurm in these vicinities (at the height of 1760 m). No boulder material is found rather below of the Adishchala gorge, and above, the gorge has a shape of a modified trough up to the village of Adishi. Starting from this village, the gorge is of a trough and its right slope is covered with moraines. The highest lateral moraine ends in the area of the altitude of 1750 m. As we saw, the length of the Adishi glacier is much behind of the length of the glaciers of the Mulkhura River basin during the Wurm. And today the lengths of these glaciers are almost equal. This phenomena has only one explanation; the basins, where we have only one glacier today and they were with only one glacier even during the Wurm, there feeding areas do not change much during the firn line depression and the firn of the given size can form relatively smaller glaciers.

The **Nenskra River basin** is one of the largest in the Enguri River basin. At one glance, the Nenskra River basin is a single old cirque above the Tita site. The field surveys and analysis of the modern aerial images showed that it is divided into several independent old cirques covering the basins of individual tributaries. There are 5 old cirques singled out in the Nenskra River basin, and the massive glacier tongues flowing out of it merged and descended to the confluence of the Lakhami River as a single tongue (Fig. 6.1). Morphologically, the relevant signs as Marghi lateral moraines are clearly seen in the relief. In each singled out cirque there are several glaciers of different morphological types at present.

The study showed that the oldest glacier was formed at the headwaters of the Nenskra River. Its length was ~ 36.0 km, while the length of the Okrila glacier was ~ 13.2 km. Both glaciers descended to the village of Marhgi. At this location, as already was mentioned, a lateral moraine is well expressed, while there are signs of the terminal moraine found in the surroundings of Marghi. The Nenskra glacier tongue descended to the height of almost 650 m during the Wurm, though, hypsometrically it descends even below, what can be explained by the location of the Enguri River. It is the first large tributary of the Enguri Rriver.

The **Dolra River basin** was a single firn basin in its headwaters during the Wurm, and ~ 34.5 km long glacier descended from it into the Enguri River bed and stretched along the Enguri gorge at ~ 2 km. At that time, the glacier occupied the territory of Tskhumari community as well and was spread along the Enguri River

gorge at ~ 0.5 km. The relief contains the indisputable evidences of this idea.

Massive glaciers were developed in the tributaries of the Enguri River basin. From some gorges they emerged in the Enguri river bed and formed certain plugs.

The following glaciers blocked the Enguri River: Khalde, Mulkhura (at two locations: near the villages of Ipari and Latali), Dolra and Lailchala. There were lakes formed above these sections what is evidenced by the lacustrine sediments preserved there.

Existence of the glacier sediments of the Wurm age at different heights in the Enguri River gorge is logical and gives the real picture of the glaciation at this location. The glaciers were spread not only in the Enguri River gorge, but also in its tributaries—Nenskra (~ 590 m), Dolra (~ 1120 m), Mulkhura (~ 1200 m), and Khaldechala (~ 1760 m), which join the Enguri River at different places at different heights, and it is naturally that the traces of Wurm moraines and glaciers are found at the same heights.

Based on the conducted researches in the Enguri River basin, we can conclude that the above-mentioned method yields reliable results and can be used in the scientific field.

6.2 Western Caucasus

The Late Pleistocene glaciation in the Western Caucasus was of a mountain-valley type. The principal glaciation center was the Caucasus watershed, while the other centers were found in the Gagra, Bzipi, Chkhalta (Apkhazeti), and Kodori ranges. Despite the fact that the Western Caucasus is not high hypsometrically, the signs of the old glaciation are well preserved in the form of cirques and moraines. The lower threshold of the cirque forms is located at the altitude of ~ 1900–2100 m. This evidences that in the Late Pleistocene the firn and snow lines depression in the river basins of the Western Caucasus was ~ 1000–1200 m. The largest glaciers on the southern slope of the Caucasus descended to ~ 600–700 m. The glaciers were of a compound-valley type and they reached

~ 16–17 km in length. The largest glacier was in the Sakeni valley (with the length of ~ 25.0 km).

The river basins of the main watershed and branches of the Western Caucasus are distinguished by the sizes of glaciers and hypsometrical locations of their tongues.

Quite a strong glaciation occurred on the southern slopes of the Western Caucasus during the Wurm. Large glaciers descended to the bottom of the Kodori River and ended at different hypsometrical levels. We have already mentioned that the lower threshold of the old glacial cirques on the southern slope of the Western Caucasus was located within ~ 1900–2100 m, i.e., at the height of ~ 2000 m on average. This altitude during the Wurm was the lower limit of the nival zone. The character of glaciation in the different river basins in the Late Pleistocene can be seen in the Fig. 6.3.

The Chkhalta River is the largest right tributary of the **Kodori River**. There are different points of views on the glaciation of its basin. Our studies showed that the old glaciers ended at different hypsometric heights on the valley bottom. The Chkhalta River begins after merging the two rivers, the Adange and the Marukhi Rivers. There were two centers of glaciation in its headwaters. In the Adange River gorge the main glacier emerged from the heads of the Amtkeli River during the Wurm and ended at the height of ~ 1700 m (Table 6.3).

A massive compound-valley glacier was found in the **Marukhi River gorge**. There were three independent cirques during the Wurm here: Chvakhra, Marukhi, and Buloni; out of them, the Marukhi cirque is the largest. The glacier flowing from it was merged to Buloni glacier and in the form of a single tongue descended to the height of ~ 1450 m in the Chkhalta River gorge. By our data, the Chvakhra glacier was smaller. It merged with Marukhi glacier and ended in ~ 1.0–1.5 km.

The same is proven by the morphological analysis of the moraines located on the right slope of the gorge. The glacier descending from the Ertsakho massif did not merge with the Marukhi glacier tongue.

The Marukhi and Adange glaciers did not merge with each other. The morphological parameters of the relief evidenced that the Marukhi glacier entered the Adange River basin at a small distance.

The glaciers descending from the Caucasus blocked the Chkhalta River gorge at three points in its lower sections. At the altitude of ~ 1000–1050 m, the Chkhalta River was merged by the Sopruju glacier from the left side. If considering the morphological conditions of the glacier, its area did not change much during the Wurm. This is why the glacier tongue was ~ 6.6 km long during the Wurm; it descended to the bottom of the Chkhalta River gorge and stretched at about 1.0 km. This is evidenced by the existence of the moraine forms in the environs of the mineral waters on the right of the valley and a bit lower (Gobejishvili 1995).

Sopruju glacier was characterized by a vast firn and a relatively short tongue during the Wurm.

There are three separate cirques in the headwaters of the **Atsiashi River**. The glaciers flowing out of them merged with each other and emerged at the bottom the Chkhalta River gorge as a single tongue. The glacier tongue ended at the height of ~ 750 m. The field geomorphological studies of the Chkhalta and Atsiashi Rivers (Gobejishvili 1995) and analysis of the modern aerial images showed that at this location the lateral moraines streach along the both banks of the Atsiashi River. The right moraine continues on the territory of the village of Marjvena Atsgara and ends right there. One should note that the villages of Marjvena Atsgara and Martskhena Atsgara are built on a plain surface, which was originated between the lateral moraine and the slope and is built by dealluvial and prolluvial sediments. Such a location of the lateral moraines allows reconstructing the width of the Wurm glacier tongue (~ 500–600 m) and its thickness (~ 60–70 m).

The **Ptishi River basin** is very interesting in respect of old glaciation. The glacier descended from here reached the Chkhalta River gorge and reached the lowest point in the Kodori River basin. There are two cirques in the head of the Ptishi River, which were the feeding basin for the Ptishi glacier during the Wurm. The glaciers

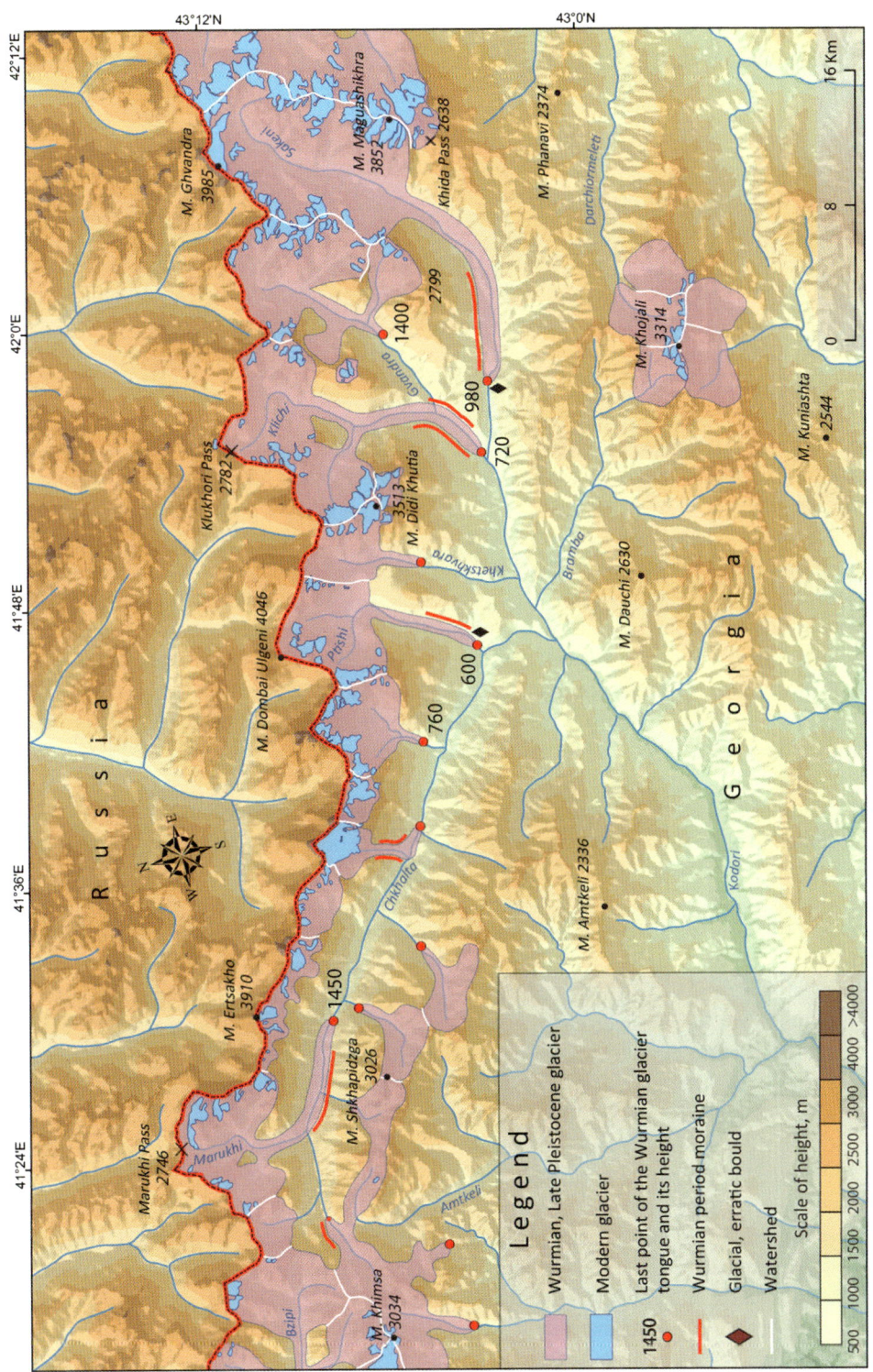

Fig. 6.3 Modern and Late Pleistocene glaciers of the Kodori River basin

Table 6.3 Morphometric parameters of the Late Pleistocene glaciers in the Kodori River basin

Name of the glacier	Old cirque area, $\sim$km^2	Length, $\sim$km	Ice-tongue height, $\sim$m
Marukhi	36.0	17.3	1450.0
Sopruju	7.7	6.7	1000.0
Adeba	8.2	7.2	900.0
Atsiashi	20.8	8.7	760.0
Ptishi	27.0	14.0	600.0
Khetskvara	12.7	7.0	1200.0
Klichi	59.0	20.0	720.0
Ghvandra	17.5	11.5	1400.0
Sakeni	89.0	25.0	980.0
Adange	8.0	7.5	1700.0

flowing out from these cirques merged with each other. In the Wurm, the length of the Ptishi glacier was $\sim$14.0 km and it descended to the height of $\sim$600 m. This was the lowest point a glacier ever reached on the territory of Georgia in the Late Pleistocene. During the Wurm, the Ptishi glacier reached the Chkhalta River gorge, and the relief shows relevant evidence: there is a fragment of the lateral moraine on the slope of the gorge opposite to the village of Ptishi, which is built with the glacial material transported from the head of the Ptishi River (Gobejishvili 1995). There are also erratic boulders on the slope scattered. Dombai-Ulgen, the highest peak of the Western Caucasus (4046 m), is located in the Ptishi River basin.

The length of the Wurm glacier in the **Khetskvara River basin** reached $\sim$7.0 km and ended at the height of $\sim$1200 m. It should be noted that there are some small cirques developed in the Khetskvara River basin. Particularly well preserved are the glacial forms on the western slope of the Khutia range. A strong flow descended from the basin of the Khutia glacier, while the other glaciers were its tributaries. The valley is of a trough shape above $\sim$1200 m. The lower section of the valley is strongly transformed due to erosion.

During the Wurm, in the Kodori River basin a large glacier flowed out from the **Klichi River gorge**. There are two clearly shaped cirques in

this gorge: one is found in the headwaters of the Klichi River, which covers the heads of the Nakhara River, and another one—in the Achapara River gorge, its right tributary. The glacier tongues flowing from the both cirques merged and descended to the Sakeni River confluence as a single tongue. Despite the fact that the lengths of the glaciers at this location are different, they ended at the same height of $\sim$720 m. The length of the Klichi glacier was $\sim$20.0 km during the Wurm, and that of the Achapara glacier was $\sim$16.0 km.

The Klichi glacier basin and its glaciers are thoroughly studied by R. Gobejishvili. The truthfulness of the lengths calculated by the method above is also evidenced by the geomorphological data. In particular, there are glacial sediments on the territories of the villages of Gentsvishi, Omarishara, and Ghvandra as moraines and erratic boulders; in addition, the whole length of the Klichi River gorge has a trough shape, though it is modified in some places due to the erosion processes. Morphologically, all these glaciers were of multi-chamber during the Wurm and their feeding basins were also quite large.

The **Ghvandra River** flows into the Klichi River from the left side and is the source of the Kodori River. The works of a number of scientists note that the Ghvandra glacier merged with the Klichi glacier. Our data do not agree with this point of view. There are some small cirques in the Ghvandra River basin. Their morphological and morphometric analysis showed that the tongues of the glaciers flowing out of the three cirques in the heads of the Ghvandra River basin merged with each other to form a single tongue and descended to the altitude of ~1400 m, below the Mindora site (in the environs of the mineral waters). There are some small cirques in the Ghvaghva and Klichi ranges. The glaciers descending to this point did not reach the Ghvandra glacier during the Wurm. Even if it merged the main glacier, it would never play any role in its existence. The length of the Ghvandra glacier was ~11.5 km. The Ghvandra River gorge is of erosive shape from the mineral waters to the Klichi River gorge.

A strong glacier was formed in the heads of the **Sakeni River**. Even today and during the Wurm, the principal glaciation center was the north-western slope of the highly elevated Kharikhra range. The glaciers descending from this slope, together with the other glaciers, flowed as a massive single flow in the Sakeni River gorge and ended in the environs of the village of Sakeni at the altitude of ~980 m. According to our calculations, the length of the glacier was ~25.0 km.

We would also like to note that there is a certain inconvenience when measuring the old glaciers. It is hard to identify the modern glacier basin playing the decisive role in the dynamics of the Sakeni old glacier. If measuring this length from the basins of Sakeni, Memuli or Chepara glaciers, the end of the paleo-glacier tongue would be located at different heights, though the distance between those heights is not big.

The glacier ending at the village of Sakeni in the Late Pleistocene is proven by the fluvial-glacial terraces developed in the environs of the village of Sakeni (Tsereteli 1966).

The glaciers in the Late Pleistocene were developed not only on the southern slope of the Caucasus range, but also over its main branches of Gagra, Bzipi, Chkhalta (Apkhazeti), and Kodori. At that time, the crests of these ranges were in the nival zone, and some massifs reached the glacial zone as well. Using the modern aerial images, allowed us to identify the old glacial

forms in each range and define the nature of glaciation by analyzing them.

There are no modern glaciers in the **Gagra range**, but the morphological and morphometric analysis of the relief evidences that there were small glaciers here in the Late Pleistocene. The cirques in the Arabika massif are well preserved (Maruashvili 1956). Their lower threshold ends at the altitude of ~2000 m. Generally, the crest portion of the Gagra range was located in the nival zone in the Late Pleistocene and due to abundant atmospheric precipitations small cirque and valley glaciers were developed here. The analysis of the cirques shows that the length of the glaciers did not exceed ~1–3 km.

The trace of old glaciation is clearly seen in the **Bzipi range**, particularly, in its eastern part, known as Chedimi range. The Mount Khimsa massif with its maximum height of 3000 m and with the small modern glaciers is located here. The old glacial forms—the cirques and moraines are particularly well developed on the northern slope of the range. The lower threshold of the cirques is at ~2000 m. The analysis of the glacial forms shows that mostly cirque and ~1–3 km long valley glaciers were developed in the Bzipi range, and the length of the glaciers on the northern slope of the Chedimi range reached ~6–8 km and reached down almost the bottom of the Bzipi River gorge. At the same time, the Mount Khimsa was in the glacial zone and a glacial cap similar to the Mount Tetnuldi was developed in it.

During the Wurm, there were four glaciers in the **Chkhalta (Apkhazeti) range**: one—in the Amtkeli valley and other three—on the northern slope of Shkhapidzga massif. The morphological analysis of the Chkhalta range relief and detailed decoding of the aerial images allowed thoroughly distinguishing the glacial forms—the cirques and moraines, which, in our opinion, were formed under the action of the glaciers of the Late Pleistocene. The Wurm glaciation of the Chkhalta range with its morphological and morphometric signs, were almost similar to the one presently found in the watershed of the Western Caucasus, from Marukhi to Klukhori Pass.

In the Late Pleistocene, quite massive glaciers were developed in the river basins of Shoudidi and Sibista and Shkhapidzga massif. The glacier tongues formed here almost reached down the Chkhalta bottom. Particularly interesting was the glacier developed in the heads of the Amtkeli River. In the Upper Early Pleistocene, the heads of the river Amkteli belonged to the river Adange basin. In the Upper Late Pleistocene, there was a massive valley glacier developed here with its tongue hanged toward the Adange valley on the one hand (most likely, the main flow flew northwards) and toward the Amtkeli valley, on the other hand. Amtkeli glacier merged with the Adange glacier (which descended from Bzipi range) and as a single tongue ended at the altitude of ~1700 m. The length of the Amtkeli glacier was ~7–8 km.

The length of the valley glaciers of the Wurm period on the northern slope of the Chkhalta range was ~4–6 km, and the length of the glaciers on the southern slope did not exceed ~1–3 km. Cirque glaciers created the glaciation background here.

In the **Kodori range** the Late Pleistocene glaciers were well developed. One can baldly state this by considering the glacial cirques and moraines well preserved in the relief. The glaciers descending from the Kodori range fed the rivers of Bramba (tributary of the Kodori River), Darchiormeleti and Larikvakva (tributaries of the Enguri River) and Ghalidzga. Large glaciers were developed in the environs of the Mount Khojali. Khojali itself was in the glacial zone and the glacial cap formed in its massif fed the glaciers of different expositions.

The largest glaciers were formed in the river gorges of Bramba and Darchiormeleti. The length of the glaciers here was ~6–8 km and the glaciers ended at the altitude of ~1600–1700 m. As L. Maruashvili noted (1956), there are many glacial lakes in the Kodori randge with the lithological factor playing an important role in their origination together with the glacier. The lower threshold of the cirque forms of the Late Pleistocene is located at the altitude of ~2100–2200 m.

6.3 Central Caucasus

The high hypsometrical location of the Central Caucasus, the morphological peculiarities of the relief, and climatic conditions determine the development of intense glaciation in the Late Pleistocene and Holocene. The lower threshold of the cirques formed by the glaciers of the Late Pleistocene is located at different hypsometrical heights. Despite this, it is well seen that the height of the lower threshold of the cirques in the Central Caucasus increases from ~2100 m (the Nenskra River basin) to ~2500 m (the Tergi River basin) from the west to the east. The main glaciation center in the Late Pleistocene was the watershed of the Caucasus range. There were quite large glaciers in the Svaneti range, while the relatively small valley and cirque glaciers were formed in the other branches of the Caucasus (Samegrelo, Lechkhumi, Shoda-Kedela, Racha, Germukhi, and Kharuli). Due to the hypsometric conditions of the branches, there was a glacial zone developed here as the individual areas. On the other hand, the entire crest zone was included in the nival zone.

Among the river basins located on the southern slope of the Central Caucasus, the Enguri River basin was distinguished by the intense glaciation. Above, we have described the glaciers of the Enguri River basin; therefore, we will briefly consider the glaciers of the Late Pleistocene below.

Our studies showed that in the Enguri River gorge, especially the glaciers descending from the basins of the right tributaries blocked the Enguri River gorge and sometimes followed it for 1–3 km (the rivers of Dolra, Mulkhura, Khalde and Lailchala). In other basins (the Nenskra, Nakra, Adishchala, and Laila) massive valley glaciers were developed during the Wurm, but they did not emerge in the Enguri River bed. Below, we consider the Wurm river basins developed in these basins.

There were four principal glaciation centers in the **Nenskra River basin** in the Late Pleistocene: Okrila, Tskhvandiri, Dalari, and Nenskra itself. The glaciers flowing from them merged and as a single flow ended in the vicinities of the village of Lakhami at the height of ∼ 650 m. Nenskra glacier was of a compound-valley type. The length of the main flow was ∼ 36 km and was the longest not only in the Enguri River basin, but also in the entire Georgia (Table 6.2). The lower threshold of the Wurm cirques is at ∼ 2100 m. The firn line depression was at ∼ 1000–1100 m. The glacial forms in the valley presented as moraines prove the fact that the Nenskra glacier reached Lakhami. An intensely degraded moraine hillock and erratic boulders are distributed in the territory of the village of Lakhami, while there is a well-preserved ∼ 5–6 km long lateral moraine in the territory of the village of Zeda Marghi on the left side of the Nenskra glacier. Nearby, in the territory of the village, on the slope, there is another well-preserved quite degraded moraine hillock. These moraines are divided form each other by the Marghi River. Their morphological analysis shows that they belong to the different periods of glaciation. The relative height between the bottom and the top of the Wurm moraine (at the village of Larilari) is ∼ 250–300 m. Existence of such a massive glacier at the village of Larilari is the evidence of the fact that the Nenskra glacier would reach down Lakhami, i.e., the glacier shifted by ∼ 4 or 5 km more.

Individual glaciation centers were located in the heads of the rivers of Tita, Lakhami, and Devra, where the glacier tongues ended at ∼ 1700–1800 m in Wurm.

About 20 km long glacier was developed in the **Nakra River basin** in Wurm. Its feeding glacial basin was formed in the head, in the basin of the modern glacier of Leadashti. The glacier tongue was made up of three main flows. The left flow ended relatively earlier, while the Leadashti and Shdavleri flows ended at the height of ∼ 1180 m at the village of Nakra with a single tongue (Fig. 6.1). According to the morphological signs of the relief, one can assume that the glacier was ended ∼ 2 km lower. There are several old glacial cirques decoded in the Shdavleri range giving the origination of

∼ 3–4 km long glaciers. Their tongues descended to the middle part of the slope to the height of ∼ 1650–1700 m.

The **Dolra River basin** was a strong glaciation center in the Late Upper Pleistocene. The morphological and morphometrical signs of the relief prove that there was a single massive cirque developed at this location, which was bordered by the Baki and Mazeri ranges from the south. The detailed analysis of the relief showed that the vast single cirque was divided into the three chambers here, such as Kvishi, Dolra, and Ushba. The glaciers descending from these cirques merged with each other at the different heights (Kvishi and Dolra merged at the height of ∼ 2200 m and Ushba glacier joined them at the height of ∼ 1900 m) emerged in the river Enguri valley as a single tongue and followed it across the territory of Tskhumari community. The length of the glacier was ∼ 34.5 km in the Late Pleistocene. The tongue of the Dolra glacier ended at the height of ∼ 1050 m (Fig. 6.1).

The glacial-morphological works conducted in the Dolra River basin and analysis of the aerial images evidenced that there is a clearly seen moraine hillock in the downstream of the Dolra River gorge, on the left side of the gorge, in front of the village of Dolra. The moraine hillock lies on the slope. Its relative height is ∼ 200–300 m and it is built with crystal rocks of different sizes, which are foreign to this site. There are also quite large untreated blocks found on the hillock and its slopes. The length of a continuous ice-tongue (lateral moraine) is ∼ 3.0–3.5 km, while it is degraded below and is presented as separate fragments. A clearly seen moraine hillock is found on the right side of the gorge, in the vicinities of the village of Nankhvari, which ends on the slope of the Enguri River. In the Late Pleistocene, the width of the Dolra glacier near the village of Dolra was ∼ 2.0–2.5 km and its strength was ∼ 250–300 m.

Next to the moraines of the Late Pleistocene, on the slope of the Bali range, at a relative height of ∼ 50–60 m, there is a strongly denuded moraine hillock with the dominating crystal rocks in its structure. The moraine material is spread over the territory of the village of Nankhvari, on the right slope of the Soledra River. We think that these moraines belonged to the Late Pleistocene glaciation.

Guli glacier merged with Dolra glacier in Wurm.

We have given the parameters of the paleo-glaciers of the **Mulkhura River basin** in the determination of the essence of the method. At this point, we will consider the glacial-morphological signs preserved in the relief and it indicates to strong glaciation. From the Lekhziri glacier to the village of Latali, the Mulkhura gorge is of a trough shape with the lateral moraines and trough bottom preserved on the slope. It is true that the bottom of the trough is transformed due to the erosive action of the river, but it is still preserved continuously on the right side of the gorge. The morphometrical parameters of the glacier of the Wurm period can be easily reconstructed based on these forms.

A whole spectrum of the lateral moraines is presented at the Mestia Airport, upward the slope to the Gvalda site. There are two moraine hillocks in the vicinities of Gvalda. The lower moraine hillock is located at the altitude of ~1900 m. It is well preserved in the relief and is adjoined by old Tskheki moraines. Another moraine is located higher and is intensely denuded. We think that these moraines belong to the two glaciations of the Late Pleistocene. The well-preserved moraine of them is of a Wurm age. There are also two moraine hillocks at the top of the village of Laghami, at the height of ~1900 m, on the right slope of the gorge. One of them is clearly seen in the relief, while another is intensely denuded. On the left slope of the gorge (in the section of the Gvalda Airport), in the relief, below the Wurm moraine, there are four moraine steps or hillocks seen clearly. Their origination is associated with the stade glaciation of the Late Pleistocene and Holocene periods. The morphological analysis of the glacial forms in the Mestiachala gorge shows that during the Wurm the Lekhziri glacier was ~3.0 km wide and its strength was ~450 m in the vicinities of the Airport.

The trace of the Late Pleistocene and Holocene glaciation is clearly seen on the northern slope of the Zagari range, on the left side of the Mulkhura River gorge. A moraine hillock follows along the slope at the height of ~1700–1800 m, which is considered of the Wurm age by all researchers of early times. The moraine hillock follows the slope at ~6–7 km, then its trace becomes obliterated due to the exodynamical processes and it appears again in the vicinities of the Ughviri Pass (Fig. 6.4). Here, the moraine hillock arches, moves to the southern slope of the Zagari range and spreads in the territory of the village of Tsvirmi. A corresponding moraine hillock is found on the other side of the Ughviri Pass, which starts on the northern slope of the Adishura–Mulkhura watershed and moves into the Enguri River gorge. There is its continuation to the left side of the Enguri gorge in the vicinities of the village of Zegani. Such a distribution of the moraines evidences that during the Wurm a certain flow of the Mulkhura glacier (perhaps, the Nageba glacier) overflowed into the Enguri River gorge through the Ughviri Pass. The flow of the glacier was spread both, upward the Enguri River gorge at about 1.0 km and downward the gorge at about 2–3 km. There are two sets of the lateral moraines inside the lateral moraines of the Wurm age in the Ughviri Pass, which arch to the south of the pass and end in the vicinities of the village of Bogreshi. These moraines are originated because of the stade glaciation of the Late Pleistocene and Holocene periods.

There are intensely denuded moraine sediments in the vicinities of the village of Keshkili, on the crest of the Zagari range, which are located ~30–40 m higher than the Wurm moraine. These glacial sediments belong to the Middle Late Pleistocene. In this period, there was an extremely cold snap on the Earth and the temperature fell by ~8–10 °C (Gobejishvili 1995).

On the right slope of the Mulkhura River gorge, from Mestia to Latali, there are only fragments of the lateral moraines preserved. They are best presented to the left side of the Tvikhulderi River as a moraine series of a minor

Fig. 6.4 Ughviri Pass and moraine hillock on the left slope of the Mulkhura River gorge (*photo by* L. Tielidze)

length. Shaping an abrupt bend by the left tributary of the Tvikhulderi River is the result of the
presence of the lateral moraines. The height of the
moraine from the trough bottom is ∼ 100–150 m.

All the above-listed moraine signs of the relief
evidence that the Mulkhura glacier was outcropped in the Enguri River bed and was in
passive contact with the Wurm glacier of Lailchala, which also was outcropped in the Enguri
River bed.

In respect of old glaciation, the highest section
of the Central Caucasus from the Mount Gistola
to the Mount Namkvani is very interesting.
Despite the fact, that there are the peaks with the
height of over 5000 m here, no massive glaciers
were here during the Wurm, which is stipulated
by the morphological conditions of the relief.
Today, there are three valley glaciers here (Adishi, Khalde and Shkhara). During the Wurm the
morphological type of these glaciers did not
change. The length of the glaciers in the Late
Pleistocene was ∼ 17–19 km. The tongues of the
glaciers ended at ∼ 1650–1900 m. The Wurm
glaciers descended to the mentioned altitude is
evidenced by the trough shape of the valley and
distribution of the glacial sediments at this
location in the form of moraines.

The **Adishchala River gorge** is of a modified
trough shape. Its gorge is asymmetrical and there

is a glacial material found on the bedrocks of the
right slope of the gorge. From the headwater to
the village of Adishi, the trough bottom, slope
and sides are clearly seen. The erosive processes
on the slopes are developed very slightly, and the
material deposited by the rocky glaciers play an
important role in it. The left slope of the gorge is
strongly inclined and fragmented and is subjected
to the intense erosive action of surface waters.
Mudflow phenomena are frequent. Because of the
erosive processes, no glacial sediments of the
Wurm period are found on the southern slope.

One must note an interesting fact in the
vicinities of the village of Adishi. On the right
side of the gorge, at the top of the village of
Adishi, there are two moraine hillocks on the
slope. The lower hillock is slightly denuded and
is well expressed morphologically. It is cut down
at ∼ 0.9–1.0 km from Adishi. The strength of the
glaciers was up to ∼ 150 m. ∼ 30–40 m above
it, there is a second intensely denuded moraine
hillock. We must note that Adishchala trough is
seen as a step in the relief at the end of the village. Differential tectonic movement stipulates
the presence of the step. The fault line is of a
general Caucasian direction.

The Adishchala River gorge has an erosive-
tectonic shape below the village of Adishi. The
glacial sediments and erratic boulders are

massively spread in the vicinities of Larnakaleri at the height of ∼1700 m. Below that, the erratic boulders are found at the altitude of ∼1600 m, in a couple of kilometers from the village of Bogreshi. In our opinion, these boulders were fallen from the right slope, from the vicinities of Ughviri in particular.

The **Khaldechala River gorge** is of a trough shape along all its length. The length of the glacier in the Late Pleistocene was ∼19.2 km. It was outcropped in the Enguri River gorge and was spread there at ∼4 km. The glacier ended in the vicinities of the village of Vichnashi at the altitude of ∼1650 m (Fig. 6.1). During the Wurm, the glacial flow from the firn of the Khalde glacier moved to the heads of the Enguri River, in particular, into the Nakarvali River gorge, the right tributary of the Enguri River. This is evidenced by the existence of the lateral moraines there and morphology of the gorge.

During the Wurm, in the **Enguri River headwaters**, the Shkhara-Namkvani glaciers were joined and as a single tongue, they ended at the altitude of ∼1900 m (Fig. 6.1) after the village of Murkmeli. The length of the glacier was ∼17.3 km. The trace of Wurm glaciation is clearly seen in the relief. The lateral moraine on the left side of the gorge near the village of Murkmeli gradually lowers hypsometrically and merges with a terminal moraine, which is preserved as a fragment. The direction of the hydrographic network in the vicinities of Ushguli, on the left side of the gorge, is determined by the location of the lateral moraines found here. This fact evidences that the glacier entered the Kvishara River gorge at ∼0.5 km as well.

The basins of the **Rioni River and its tributaries** are located on the southern slope of the middle section of the Central Caucasus. In respect of old glaciation, it is both, very interesting and problematic. The scientists have quite different opinions about the identification of the boundaries of the Wurm glaciation (Vardaniants 1930, 1937; Maruashvili 1956; Kovalev 1961; Astakhov 1973; Dondua 1959; Tsereteli 1966; Scherbakova 1971; Khazaradze 1968; Gobejishvili 1995, and others).

Our investigations allow expressing a different opinion. The southern slope of the Caucasus watershed from the Mount Namkvani to the Mount Phasismta is a part of the Tskhenistskali River basin. There were three glaciation centers here during the Late Pleistocene—in the heads of the rivers of Koruldashi, Zeskho, and Tskhenistskali itself.

There was quite a strong glacier in the **Koruldashi River basin** during the Wurm. It was a valley glacier with the length of ∼11.7 km ending in the vicinities of the village of Tsana at the altitude of ∼1600 m (Table 6.4). The glacier ending at the village of Tsana is evidenced by a number of doubtless glacial-morphological signs. The whole length of the Koruldashi gorge is of a trough shape from the head up to Tsana. Below Tsana, at ∼60–70 m above the riverbed, the fluvial-glacial material with slightly processed block is bedded on the Lias pack evidencing its glacial origination.

The Zeskho glacier was of ∼7.5 km long during the Wurm and ended at the height of ∼1600 m. The glacier was fed from the slope of the Mount Tsurungala. There was a ∼4–5 km long independent glacier in the Zeskho River heads. One should note that during the Wurm, the glaciers of Koruldashi and Zeskho ended by ∼2–3 km above the confluence of the rivers. The downstream of the Zeskho gorge is erosive.

In the Late Pleistocene, in the Tskhenistskali River heads, there was one valley glacier with its tongue in the vicinities of Lapuri location ending at the height of ∼1800 m. The length of the glacier was ∼9 km.

The Caucasus watershed range from the Mount Pasis-Mta to the Mamisoni Pass is known as a Racha Caucasus. It includes the Rioni River heads and is very interesting in respect of old glaciation. All previous researches considered that all the glaciers found here merged with each other during the Wurm and they had a single glacial tongue. By our investigations, they give a different picture. During the Wurm, all glacial basins were isolated. The tongues of some of them outcropped only onto the bottom of the

Table 6.4 Morphometric indicators of the Late Pleistocene glaciers in the Central Caucasus

№	Name of the glacier	Old cirque area, $\sim km^2$	Length, $\sim km$	Height, $\sim m$
	Enguri River basin			
1	Nakra	63.0	20.0	1180
2	Tkheishi	7.0	5.6	1450
3	Kveishkhi	4.5	3.6	1600
4	Didgali	6.0	4.8	1350
5	Magana	7.5	6.0	1450
6	Khobistskali	8.0	6.4	1800
	Rioni River basin			
7	Korulashi	15.0	11.7	1600
8	Zeskho	9.0	7.5	1600
9	Shari	11.5	9.0	1800
10	Edena	16.5	14.1	1650
11	Zopkhito	21.5	17.4	1500
12	Kirtisho	42.0	20.5	1300
13	Notsara	14.8	12.6	1280
14	Boko	20.9	18.0	1100
15	Buba	40.5	23.0	1050
16	Chanchakhi	13.1	11.0	1750
17	Gharula	17.5	14.0	1250
18	Jejora	19.8	17.5	1600
19	Latashuri	11.0	8.5	1500
20	Sokhortuli	7.5	6.0	1600
21	Ghobishuri	8.5	7.0	1800
22	Shodura	5.4	4.0	1600
	Liakhvi River basin			
23	Zakara	8.5	7.0	1900
24	Kveshelta	10.0	8.0	1800
25	Jomagi	11.2	9.1	1550
26	Sba	9.2	7.5	1800
27	Cheliata	8.0	6.5	1850
28	Kalasani	11.0	9.0	1800
	Tergi River basin			
29	Devdoraki	38.5	14.2	1220
30	Gergeti	21.0	17.0	1550
31	Mna	23.0	18.0	1950
32	Suatisi	32.0	15.0	2150
33	Tergi head	20.0	10.0	2270

Rioni River gorge and blocked it. The glacier tongues ended at different hypsometrical altitudes (Fig. 6.5).

During the Wurm, in the heads of the Rioni River, there were two huge glaciers—Edena and Zopkhito. During the Wurm a strong glacier with the length of ~17.4 km descended into the **Zopkhitura River gorge**. Its tongue outcropped in the Rioni River gorge and ended at Khopito location, at the height of ~1500 m (Fig. 6.6). There are the moraine sediment outcrops with the granite block under the Khopito debris cone. The Wurm cirque is clearly seen in the relief and it is fixed at the height of ~2200 m, i.e., the firn line ran at the height of ~2200 m (Gobejishvili et al. 2012).

Near the village of Ghebi the **Chveshura River** joins the Rioni River from the left side. By our data, the glaciers formed in this basin during the Wurm merged with each other and emerged into the Rioni River bed as a single tongue.

The length of the glacier located in the Chveshura River gorge was ~20.5 km. Its ice-tongue was ended at the height of ~1300 m. This was a compound-valley glacier. The glacial forms are well preserved in the Chveshura River gorge. There are two rows of lateral moraines in the vicinities of the site called Jojokheta. The lower moraine hillock follows the left slope of the Chveshura River gorge and extends at ~1.0–2.0 km in the Rioni River gorge. The upper hillock is denuded and is survived as a fragment. Both moraines are of the Late Pleistocene age, and the preserved hillock must belong to the Wurm glaciation. In the territory of the village of Ghebi, together with the moraine material, there are many erratic boulders (Fig. 6.7). Generally, the Chveshura River gorge is of a trough shape modified by the erosive processes.

The glacial material is absent in the gorge from the village of Ghebi to the Saglolo site, what indicates that the Chveshura glacier was not merged with the Chanchakhi glaciers.

During the Wurm there was a strong glaciation center in the **Chanchakhi River gorge** in the Rioni River basin; the glacier was of a compound-valley type. Boko, Buba, and Tbilisa glaciers merged with each other and as a single tongue descended to the Tsidro fortress in the Rioni River gorge. It should be noted that the glaciers flowing from the Kedela range merged with the main glacier. This is well evidenced by the hanging troughs survived in the relief. There are clear signs of glacier action at the Saglolo site and a bit lower as numerous granite blocks, both, in the modern bed and on the slopes. This fact is described in the works of many scientists who, following this fact, assume that the glaciers of the Rioni River basin descended to these locations as a single tongue. Gobejishvili (1995) does not share this view. He thinks that there was no single glacier in the heads of the Rioni River, and only the Buba-Boko glacier fed from a massive glacial plateau of Karaugomi descended to these places. The length of the glacier was ~23 km and its tongue ended at the height of ~1050 m (Fig. 6.5).

Some researchers of the Caucasus consider that the glaciers in the Rioni River basin reached down the city of Oni (Astakhov 1973). Their opinion is based on the analysis of the fluvial-glacial material near the city of Oni. It is true that this point view has the opponents, but the presence of the erratic granite boulders here makes it difficult to find a single solution of this question.

An awful mudflow on the Rioni River in August 15 of 1989 stripped a huge granite boulder, which was covered with floodplain vegetation before that. The locals call it a "Feast stone." The high water level in the Rioni River (~2–3 m along the whole riverbed) destructed the floodplain vegetation around the boulder, cleared the granite boulder and makes it clearly visible. The boulder is located in the riverbed on the right side of the Rioni River, ~150 m above the confluence of the rivers of Rioni and Gharula. Its width is ~14 m and its length is ~12 m. The height of the visible part is ~3.5–4 m, and perhaps the same height is buried into the bed. The perimeter of the boulder is ~36 m and its volume is approximately 1000 m^3. The weight of the erratic boulder is more than 2000 tons

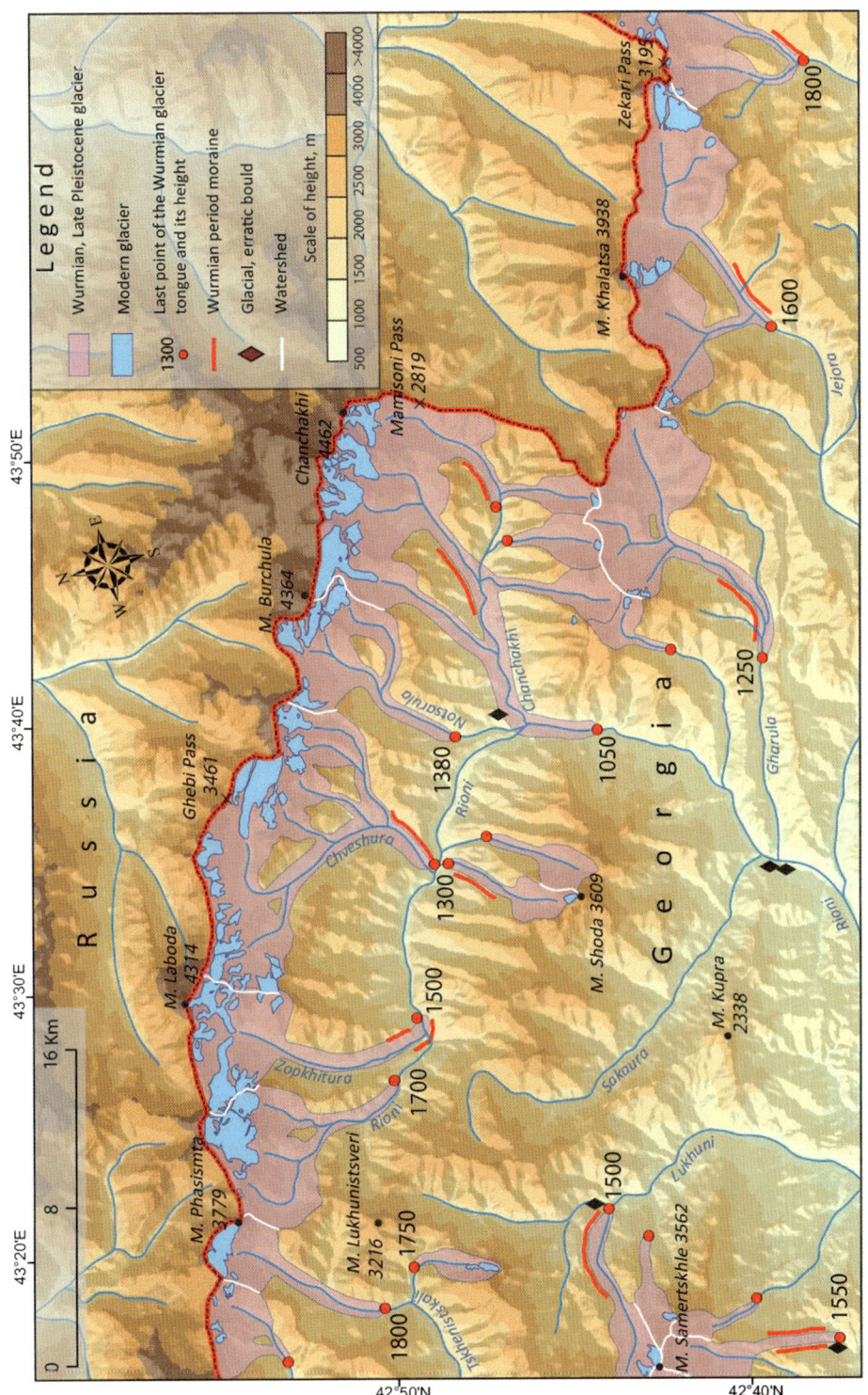

Fig. 6.5   Modern and Late Pleistocene glaciers in the Rioni River basin

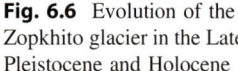

Fig. 6.6 Evolution of the Zopkhito glacier in the Late Pleistocene and Holocene

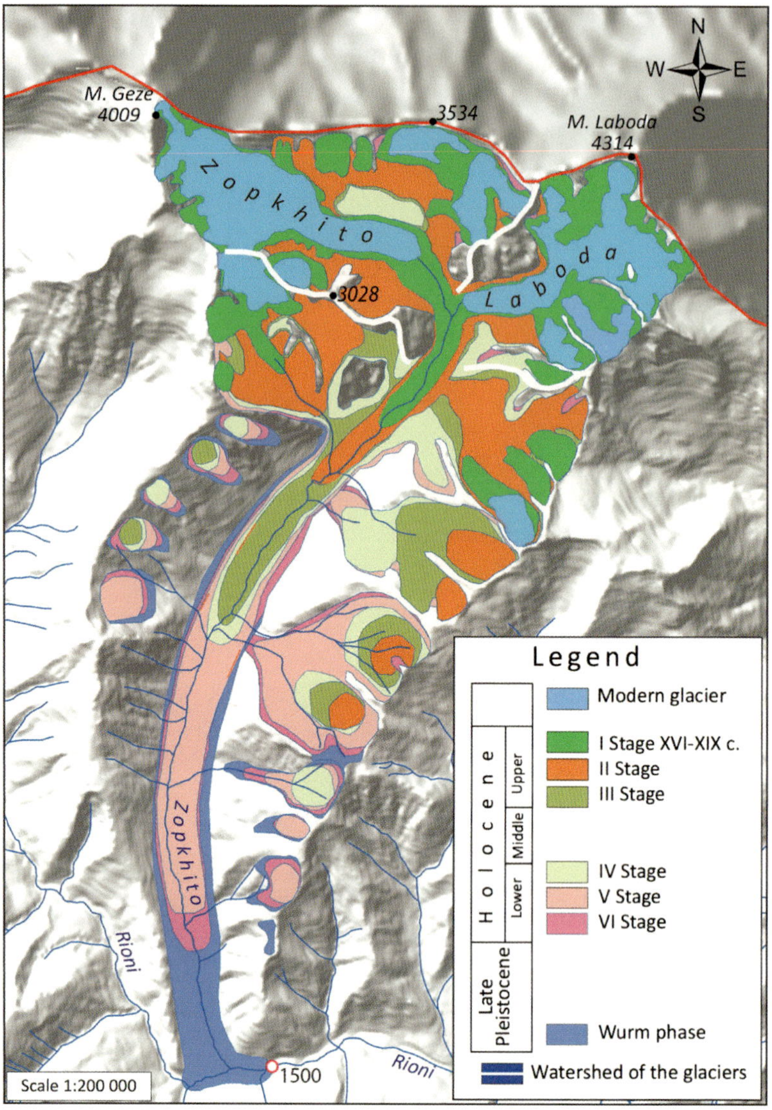

(Fig. 6.8). The petrographic study of the block showed that it is a microcline granite, and the Racha Caucasus is built by the rocks of the given structure and content (Gobejishvili 1995).

Another erratic boulder found in the Rioni River bed, near the confluence with the river Sakaura, is a biotitic plagiogranite. Similar rocks build the relief of the Boko and Buba basins in the Chanchakhi River gorge.

The length of the erratic boulder is ~10 m, its width is ~5 m, its height—~5 m, its volume is ~200 m³ and its weight is ~500 tons.

The complete study of the erratic boulders demonstrated that they are unfamiliar to these places and are brought here from the Caucasus range. As for the kind of force capable of bringing these boulders here, as already mentioned, there are two major explanations to this question; one of them suggests that the glacier brought the boulders from the Caucasus range to Saglolo, and then the glacial mudflow (strong flooding) formed in the Rioni River gorge brought them to the city of Oni. Bringing the granite boulders by the mentioned means is much

Fig. 6.7 Erratic boulder in the Chveshura River gorge in the territory of the Ghebi village (*photo by* L. Tielidze)

Fig. 6.8 Erratic boulder in the Rioni River gorge (*photo by* R. Gobejishvili)

doubtful due to the following reasons: the morphological analysis of the gorge and bed along the Oni-Saglolo section showed that the gradient of the Rioni riverbed in fact has not been changed for the last ∼20 thousand years. Both, today and earlier, the inclination of the Rioni riverbed has been ∼15–20 meters per kilometer. In the conditions of such a minor riverbed inclination, any kind of flooding could not roll down such huge boulders. Another argument is the flooding on the Rioni River in 1995 and 2005, when the Rioni demolished bridges and roads and washed out old terraces, but could not even move these boulders. Another view suggests that these boulders were shifted here by a glacier. One group of researchers considers that a strong glacier tongue in the Rioni River basin descended to the city of Oni. They based their opinion on the fact of the presence of the erratic boulders here. The complete glacial-morphological study of the Rioni River headwaters showed that during the last strong glaciation (∼20–25 thousand years ago) the glacier tongues in the Rioni basin descended to Tsidro fortress (below Saglolo). Below to this location to Oni, there are no traces of a long-term action of any glacier. This means that the mechanism of transporting the granite boulders to this location is unclear.

It is true that if we would like to understand the mechanism of transporting, we should use the comparison method (consider the action of the modern glaciers in the past). In recent years, the information gained from different mountainous regions of the world and old literary sources have made it clear that the modern glaciers (in some cases) are characterized by a pulsating movement. At this time, the regime of a glacier dynamics is harshly disturbed followed by the redistribution of the energy and mass within the glacier; as a result, the topographic surface of the glacier changes, but its total balance is the same. A sharp movement of a glacier body from the firn basin toward the tongue causes a rapid movement of the glacier tongue forward. At present, it is known that the end of the glacier tongue of this type will advance by ∼10–15 km per 2 or 3 months. The glacier tongue descended to the low altitudes melts intensely and rapidly retreats.

It should be noted that the pulsating (furious) glacier tongues when moving toward the valley, leave the large blocks on their surfaces at the locations where the tongues stop moving. They do not form any other forms in the relief. Presumably the Rioni River basin was characterized by the same conditions ∼20 thousand years ago. The strong glacier tongue descended below Saglolo advanced sharply and reached the city of Oni. It left the granite boulders on its surface in the vicinities of Oni. All this time, the squared granite boulders were subject to the action of the Rioni River water. The river water abraded and rounded them, and today they are of such form.

Hypsometrically, the eastern section of the Central Caucasus is much lower than its western part. Some peaks here are more than 3700 m in height (Saukhokhi, Khalatsa, Brutsabdzeli, Zekari, Zilgakhokhi, and Laghztsiti) and there are modern glaciers developed around them. This section of the range from Kozi to Vatsisparsi is known as the Dvaleti Caucasus, while its eastern part is called the Mtiuleti Caucasus.

The Gharula and Jejora basins of the left tributaries of the Rioni River and the Liakvi River basin are located on the southern slope of the Dvaleti range.

In the **Gharula River basin** there was a strong glacier in the Late Pleistocene, which was fed from a four-chamber cirque and descended to the village of Kvemo Kvazha at the height of ∼1250–1300 m as a single tongue. The length of the glacier was ∼14 km (Fig. 6.5). We note that there are clearly seen lateral moraines on the territory of the village of Kvemo Kvazha. The terminal moraines are washed out (Dondua 1959; Tsereteli 1959). The trace of stade glaciers is seen in the head of the valley. During the Wurm, the lower threshold of the cirque was at the height of ∼2300–2400 m evidencing that in the Wurm the firn line depression was at ∼1000–1200 m.

The Wurm glaciation of the **Jejora River basin** is studied by many scientists (Dondua 1959; Tsereteli 1966). They have different opinions about the identification of the lower limit of the Wurm glacier. Our investigations have made it clear that there were two huge valley glaciers descending to the Jejora River basin and they merged at the village of Kveselta (at ∼1750 m) and flowed below the village of Kobeti as a single tongue. The maximum length of the glaciers was ∼17.5 km, with one flow ending at the village of Kobeti (at ∼1600 m) and another flow ending above the village of Nakrepa, at the height of ∼1550 m. The lower threshold of the Wurm cirques is at the height of ∼2200–2300 m in the Jejora River basin.

The Caucasus range within the **Liakhvi River basin** is characterized by low hypsometrical signs. Both, modern and old glaciers are widely spread in its western part, including the Mount Zekara and a part of the Mount Brutsabdzeli and Liakvhi-Tergi watershed in the east. In the Wurm, there were two separate glaciers in the heads of the Patsa River. Their length was ∼7–8 km and their tongues ended near to each other, at the height of ∼1800 m. During the Wurm, the tongues of the both glaciers made the tongue of the Jomagi glacier. The length of the glacier was ∼9.1 km and it ended at ∼1550 m above sea level.

During the Wurm the glaciers in the tributaries of the Liakhvi River (the rivers of Sba, Cheliata, and Kalasani) ended in the vicinities of the villages of Shua Sba and Zemo Cheliata at the height of ~ 1750–1850 m. In a glacial-geomorphological point of view, the Kalasani River gorge is very interesting. Glacier Laghztsiti is located there, and there is a well-preserved cirque in the heads of the gorge with its lower threshold at ~ 2400–2500 m. By our investigations, during the Wurm, there were Wurm-type glaciers there with the length of ~ 12 km. The lower part of the gorge is filled with volcanic lavas, and the glacial signs are almost cleared out. During the Wurm, the glacier descended into the territory of the village of Edisa and may be, even lower. The trace of the Wurm glaciation was swept first by Akhubati lava flow and then by Khodza lava flow. Identification of the age of Akhubati lava current is a bit difficult. It is a fact that this lava flow essentially changed the morphology of the Liakhvi gorge above the village of Akhubati (Gobejishvili 1995).

An easternmost part of the southern slope of the Central Caucasus belongs to the **Keli volcanic plateau** and the basins associated to it.

In a glacial-geomorphological and geological point of view, the Keli plateau and adjacent ranges are studied quite well by L. Maruashvili, E. Milanovskiy, N. Skhirtladze, D. Tsereteli, N. Dzotsenidze, and others. The approximate age of the volcanoes and lava flows is established by means of correlation with fluvio-glacial sediments. An opinion about the presence of a glacial shield (range) on the Keli plateau during the Wurm was expressed.

The investigations conducted during the recent years, particularly using the aerial images, have demonstrated that some questions are disputable and need further clarification.

The trace of the Late Pleistocene glaciation is survived quite clearly on Keli plateau and adjacent ranges as cirques, troughs, and moraines. However, sometimes, under the action of the volcanoes, these forms are modified. The cirques preserved on the southern slope of the Caucasus and on Kharuli rage are small in size (~ 2–4 km); cirque, and ~ 2–4 km long valley

glaciers were common there. As for the northern slope of the Caucasus, the length of the glaciers was ~ 6–8 km and there feeding areas (cirques) reached ~ 8–10 km². It should be noted, that there are some small-sized cirque forms developed within the large cirques.

The cirques developed in the Mepiskalo volcanic relief are interesting. Their area is ~ 2–4 km². During the Wurm, the valley glaciers descended into the Tetri Aragvi River gorge. Development of the glacial forms of the Late Pleistocene in the volcanic relief evidences that the volcanic relief was formed either in the Late Pleistocene, or earlier.

Many volcanic cones (Didi and Patara Sherkhota, eastern Khomisari, and Tsiteli Khati) are located in the glacial cirque of the Wurm age evidencing that these cones were originated in the post-Late Pleistocene age and quite fairly, some scientists belong them to the Holocene. The volcanic cones located in the Keli plateau (Narvani, Keli, Sharkhokhi, Shadilkhokhi, and others) and related to them lavas are considered to be of the Holocene age by considering the morphological signs of their origination. This is evidenced by the fact, that these cones, despite their high hypsometrical location, unlike Mepiskalo volcano, have no signs of glacial action.

The analysis of the morphological–morphometric properties of the relief of Keli plateau and location of the snow line in the Caucasus in modern times and in the Wurm demonstrated that not all area of the plateau was subjected to the impact of the glacial zone. Here, at some locations, there were small cirque and valley glaciers developed. The nival processes were very active in the past and they are active even today.

The glaciation in the Upper Pleistocene was more or less presented in the **branch-ranges of the southern slope of the Central Caucasus**. As the **branch-ranges** are of general Caucasian or lateral direction, the glaciation degree here is determined by the relief hypsometry, morphology, and distance from the Black Sea.

The glaciation in the **Svaneti range** had a sharply asymmetrical nature in the Late

Pleistocene and still has today, what in our opinion, is caused not only by exposition, but also by the morphological conditions of the relief.

The largest glaciers were Laila and Lailchala. Their tongues reached the height of ~ 1380–1100 m and their length was ~ 12–13 km. It should be noted that the Lailchala glacier outcropped on the bottom of the Enguri River (Fig. 6.1). This is evidenced by the moraine hillocks (in total, two of them) bedded in the vicinities of the church in Laila, which form steps hypsometrically and belong to different glaciations. As for the Lailchala gorge, it is of a trough shape.

Valley glaciers were found in the basins of the rivers on the northern and southern slopes of the Svaneti range (the rivers of Arshira, Lasiri, Kvishara, Skilori, Laskadura, Mukhra, and Ashkhashuri). The length of the glaciers on the northern slope was ~ 5–7 km and the lower threshold of the glacial cirques is at the height of ~ 2100–2200 m.

The length of the valley glaciers on the southern slope of the Svaneti range was ~ 3–5 km. Old glacial forms, cirques and moraines are relatively weakly preserved morphologically. However, they carry sufficient information to reconstruct the picture of the old glaciation. Large glaciers descended to the height of ~ 1700–1800 m.

The Samegrelo range is distinguished by the peculiarities of modern and old glaciations. Its highest peaks reach 3200 m. The modern glaciers of small sizes are comfortably set in the relief. They are formed at the height of ~ 2800–2900 m and the ice-tongues of some of them descend to the height of ~ 2400 m. In fact, the glaciers are originated below the nival zone (in the alpine zone) what is quite rare for the Caucasus. The major factors of glacier formation are the peculiarities of a relief and close location to the Black Sea (abundant atmospheric precipitations in the cold period). In our opinion, there were similar conditions in the Late Pleistocene. Using the modern aerial images, we were able to reconstruct the picture of the glaciation in the Late Pleistocene. The glacial cirques are well preserved, particularly in the relief built by Bajocian porphyries. The area of the glacial cirques was ~ 3–6 km^2 (Table 6.4).

The lengths of the modern glaciers and areas of the feeding basins in the Samegrelo range are almost the same. The length of the large glaciers in the Pleistocene was ~ 3–6 km. Such a length was sufficient for the glaciers to reach the lower hypsometrical level (~ 1350–1450 m). The glaciers reached the lowest hypsometrical height in the Magana River basin. In the Didgali River gorge the right tributary of the Magana River two glaciers merged with each other and ended at the height of ~ 1350 m as a single tongue. At the same height and a bit higher, there are lateral moraines survived in the relief what is an undisputable sign of the presence of glaciers. Generally, this basin is very interesting in a glacial-geomorphological point of view. The glacier formed in the head of the Magana River reached ~ 6 km in the Late Pleistocene and reached the height of ~ 1450 m, where it was merged by the Magana glacier.

During the Wurm, there were many valley and cirque glaciers found in the **Khaishura River gorge** (northern slope of the Samegrelo range) evidenced by the old cirques well preserved in the local relief. The glacier formed in the head of the Tkheishi River is worth mentioning. The length of the glacier was ~ 5.6 km. The tongue of the glacier was made up of four flows and descended to the height of ~ 1450 m. The trace of the action of the glacier tongue is morphologically well survived in the relief.

The glaciation of the Late Pleistocene was well seen in the **Khobistskali and Tekhura River basins** on the southern slope of the Samegrelo range. Many large glaciers were formed there due to the northeastern exposition of the cirque in the head of the Khobistskali River. By our data, the length of the glacier was ~ 6.4 km and it ended at the height of ~ 1800 m. According to some of the signs survived in the gorge, it is possible that this glacier would be even longer and reached the height of ~ 1700 m.

It is true that there are no glaciers in the eastern part of the Samegrelo range in the vicinities of the Mount Tsekuri, but the trace of the action of the Late Pleistocene and Holocene glaciers is well seen here. The area of the old cirques located here is ~ 1–3 km^2.

According to the conducted studies, the firn line height was ~ 2000–2100 m during the Wurm.

In the Late Pleistocene, the crest section of the Lechkhumi and Shoda-Kedela ranges was totally included in the nival-glacial zone. And some peaks, the heights of which exceeded 3300 m, included the glacial zone as well.

In respect of glaciation, there is a Chudkharo–Samertskhle porphyry massif identified in the **Lechkhumi range**. During the Wurm, quite large valley glaciers descended from this point in every direction. The conducted glacial-geomorphological works allowed us to identify the lower limit of extension of the glaciers. According to our investigations, there was an ~ 8.5 km long glacier in the river Latashuri gorge with its ice-tongue descending to the altitude of ~ 1500–1550 m. The glacier almost totally occupied the Latashuri gorge, and there are many glacial-morphological evidences of this: presence of porphyry blocks up to the Latashuri–Lukhunistskali confluence, clearly expressed lateral moraines of the Late Pleistocene and Holocene ages on the left side of the gorge, morphology of the gorge, etc. The porphyry blocks in the vicinities of the site of Kajiani are associated with the Latashuri glacier. These blocks were rolled down from the Latashuri–Kajiani watershed, where the moraine material was accumulated during the Wurm glaciation.

There were valley glaciers in the heads of the Ghobishuri and Zhrinavi Rivers. The length of the glaciers was ~ 5–7 km and they reached the height of ~ 1800–1900 m. This is evidenced by the glacial forms, in particular, by the moraines, which are well preserved in both gorges. This region was studied by Nemanishvili (1962), Tsereteli (1966), and others. Our point of views more or less coincides with their opinions.

The glaciation of the **Shoda-Kedela range** in the Wurm was asymmetrical. The glaciation was well developed on the northern slope. There were valley glaciers here with the length of ~ 1–3 km. The headwaters of the Sakaura River also belonged to the northern slope and the local glacial rivers flew directly into the Rioni River. In the Upper Pleistocene, the Sakaura River intruded the northern slope by regressive erosion and captured its present head. The gorges of the northern slope of Shoda-Kedela still maintain a trough shape today.

The trace of the old glaciation is preserved in the high-mountainous relief of the **Racha, Germukhi, and Kharuli mountain ranges**. We decoded all glacial forms and calculated the area of the glacial cirques. It turned out that the cirques are characterized by small areas—~ 1–4 km^2 in all mountain ranges. These data indicate that the maximum lengths of the glaciers in these ranges do not exceed the ~ 4.0 km. With their morphological and morphometric properties of the glacial forms, the local glaciers were of small size valley and cirque types. The lower threshold of the glacial cirques was at the height of ~ 2400–2600 m, i.e., the firn line was stretched at this height in the Wurm. A very interesting research of the old glaciation of the Germukhi range belongs to Maruashvili (1961).

6.4 Eastern Caucasus

The relief of the Eastern Caucasus with its morphological and morphometric properties is behind the Central Caucasus, resulting in, together with the climatic factors, the small scales of glaciation both, today and in the entire Upper Quaternary and Holocene. As we have seen above, the modern glaciers are associated with the individual high-elevated massifs. A similar picture was observed in the Wurm, as by our calculations, the lower limit of the glacial zone in the Wurm was at the height of ~ 3400–3600 m. The nival zone was presented at a vast area in the Eastern Caucasus, but it was not extended over the entire territory. It was spread continuously only along the Caucasus watershed ridge. As for the Wurm glaciers, they were linked with the massifs with the heights of about 3400–3600 m. We describe the old glaciation according to the individual orographic units in relationship with the river basins.

The **Tergi River head and the basins of its tributaries** are located on the northern slope of the Eastern Caucasus. Here, the northern exposition of the slope plays a certain role in the origination of both modern and Wurm glaciations. The old cirque forms here evidence that during the Wurm there were quite strong glaciers in the basins of the right tributaries of the Tergi River with their tongues outcropping into the bottom of the Truso basin (the lower threshold of the glacial cirques is at the height of ~2500–2600 m). The longest glaciers were in the head of the Tergi River and the Desistskali gorge. Their length was ~10–12 km.

In the Late Pleistocene the principal center of glaciation in the Tergi River basin was and even today is the lateral range of the Caucasus. In the Late Pleistocene the Khokhi range in the vicinities of the Jimara-Kazbegi massif was covered by a strong glacial cap. The glacier tongues descended in all directions of the massif. Large glaciers: Devdoraki, Gergeti, Mna, and Suatisi were of ~14–17 km long. They emerged on the valley bottom. The glaciers formed plugs in the valley and often promoted the formation of strong glacial mudflows. In our point of view, the glacial flow was not single in the Tergi gorge. The glacier tongues ended at the different hypsometric altitudes. The Devdoraki glacier descended to the lowest hypsometrical point. It was merged by the Chachi glacier from its left side, passed the Dariali gorge and reached Zemo Larsi. It was this glacier, which brought a huge granite boulder, so called "Ermolov boulder" to Zemo Larsi (Fig. 6.9). The glacier tongue ended at the height of ~1200 m. The Gergeti glacier was joined by Abano glacier and ended at the height of ~1550 m above sea level below the settlement of Stepantsminda as a single tongue (Fig. 6.10).

The Mna glacier had the longest ice-tongue. The glacier outcropped in the Tergi River bed by its strong tongue because of the morphology of the valley and almost reached the village of Kobi

at ~1950 m above sea level. All three tongues of the Suatisi glacier ended at ~2150 m above sea level in Truso gorge as a joint flow.

There were large glaciers in the basins of other rivers on the southern slope of the Khokhi range (the rivers of Jimara, Tepistskali, Resistkali, and Siverauti). It is true that the lengths of these glaciers in the Wurm were ~5–8 km, but some of them reached the Truso depression (Fig. 6.10).

The modern firm line in Tergi River basin is located at the height of ~3500–3600 m, and the snow line is at the height of ~4400–4500 m. During the Wurm the firn line was at the height of ~2400–2500 m and the snow line was at the height of ~3400 m. According to these data, during the Wurm the firn line depression was ~1100 m on average. The area of the nival-glacial zone in the Tergi River basin in the Wurm was ~135 km^2.

A strong glaciation center on the Caucasus watershed ridge was the **Chaukhi massif**. In the Late Pleistocene the glaciers from this point descended to all directions. The largest glacier on the southern slope of the watershed ridge was the Abudelauri glacier, which was descended into the Roshka River gorge, the right tributary of the Khevuretis Aragvi River. According to our calculations, the length of the glacier was ~11.5 km and its tongue ended at the height of ~1600 m (Table 6.5). At the point of confluence of the rivers of Roshka and Aragvi and there are erratic diabase boulders of large size above it. This evidences that during the Wurm the glacier descended to this mark or even below it. The boulders located at the village of Lelisvake belong to the glaciation of the Middle Late Pleistocene.

The Chaukhi glacier was descended to the village of Juta to the height of ~2200 m from the Chaukhi massif on the northern slope. Its length was ~5.7 km. The Chaukhi River gorge maintains the trough shape along the entire length. The lower threshold of the Wurm cirques on the Chaukhi massif is at ~2400–2500 m, while the firn line depression was of ~1100–1200 m.

The valley glaciers in the **Juta River basin** were located in the basins of the rivers (the rivers

Fig. 6.9 "Ermolov boulder", Zemo Larsi (*photo by* R. Gobejishvili)

of Kora, Javartkhokhi, Narvanistskali, and others) on the southern slope of the Kuro and Shavana ranges. The aerial images clearly show the old cirques with their lower thresholds at the height of ∼2500–2600 m. The length of the glaciers located there was ∼4–6 km. Their ice-tongues ended at the height of ∼2200–2300 m. It is true that according to our data, the glaciers were not merged with each other, but the morphological analysis of the relief in the heads of the Juta River makes us think that at this location, the glaciers merged with each other; and this point of view surely needs a detail investigation.

The glaciers on the eastern slope of the **Kuro range** and western slope of the **Shavana range** were merged with each other in the Wurm and in the head of the Khdestskali River and they fed a valley glacier of Kibesha. The length of the glacier was ∼11.9 km, and its tongue ended at the height of ∼2140 m. The upper section of Khde gorge is of a glacial genesis what is hardly true with its lower section.

The eastern slope of the **Kidegani and Arjelomi ranges** today and in the Wurm, has been the major center of glaciation in the Asa River basin. Two glaciers were descended from the Arjelomi range to the Chkhotani River basin.

Their length was ∼6.5 km and they were ended at the height of ∼1700 m. This glacier did not descend into the Asa River gorge. The largest glacier in the **Asa River gorge** was the Akhieli glacier. It had two flows, descended into the bottom of the Asa River gorge and spread below ∼1.5–2 km at the village of Akhieli. The length of the Akhieli glacier was ∼7.3 km and its tongue was ended at the height of ∼1750 m above sea level. In the head and tributaries of the Asa River, where there are no glaciers at present, small old glacial cirques have survived evidencing the presence of small cirque and valley glaciers there in Wurm. The lower threshold of the old cirques in the Asa River gorge is at the height of ∼2400–2500 m. Therefore, the firn line depression in Wurm was ∼1000–1100 m.

Khevsureti range is a watershed of the rivers of Asa and Arghuni. There are two high massifs identified there—Amghismaghali and Makhismaghali massifs. The trace of the action of the Wurm glaciers has been preserved exactly around these massifs.

There were many cirque and valley glaciers in the Asa River gorge on the western slope of the Khevsureti range. Among them, the Bisna glacier (with the length of ∼5.5 km and height of the tongue ∼2060 m) and Kogra glacier are worth

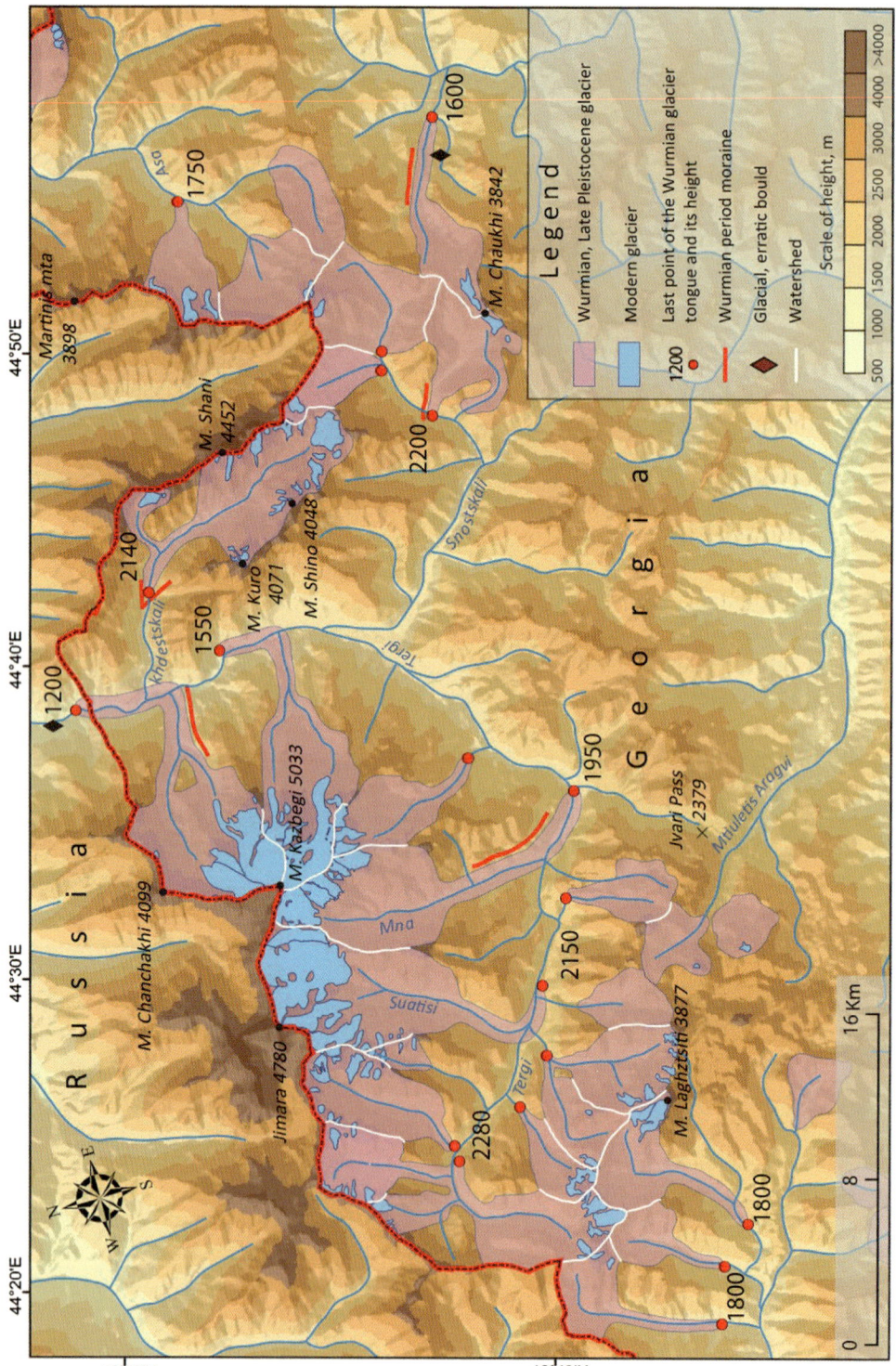

Fig. 6.10 Modern and Late Pleistocene glaciers in the Tergi River basin

Table 6.5 Morphometric parameters of the Late Pleistocene glaciers in the Eastern Caucasus

No.	Name of the glacier	Old cirque area, $\sim$km^2	Length, $\sim$km	Ice-tongue altitude, $\sim$m
	Aragvi River basin			
1	Abudelauri	14.2	11.5	1600
	Tergi River basin			
2	Chaukhi	7.0	5.7	2200
3	Khdestskali	24.0	11.9	2140
	Asa River basin			
4	Chkhotani	–	6.5	1700
5	Akhieli	–	7.3	1750
6	Bisna	–	5.5	2060
	Arghuni River basin			
7	Shatili	10.5	8.5	1800
8	Makhi	9.5	7.0	1900
9	Kamkhi	5.5	4.5	2100
	Pirikita Alazani River basin			
10	Tebulo	14.0	11.0	2400
11	Amugho	7.5	6.0	2500
12	Larovani	6.8	5.5	2560
13	Kachu	11.5	6.0	2200
14	Didkhevi	13.7	7.0	2220
15	Cheros Khevi	10.0	5.0	2140
16	Diklo (Khao)	14.5	7.0	2100

to note. Large glaciers were developed during the Wurm on the eastern slope of the Khevsureti range. We would like to mention two independent glaciers in the Shatilistskali River basin. They started in Makhismagali massif and descended to the height of $\sim$1800–1900 m. The length of the Shatili glacier was $\sim$8.5 km. The smaller glaciers were in the Giorgitsminda and Guristskali River basins.

There are several old cirques remained around the Ardotistavi massif, and there is a small cirque glacier remained in the head of the Alerdoy River. The Alerdoy River gorge is of a trough shape along $\sim$3 or 4 km, and there is a trace of old rocky glaciers in its headwater.

The lower threshold of the old cirques on the both slopes of the Khevsureti range is at the height of $\sim$2400–2500 m.

In the **Andaki River basin**-the tributary of the Arghuni River, the Wurm glaciers were developed on the western slope of the Mutso and Atsunta ranges. Atsunta range isolates the Arghuni River basin from the Pirikita Alazani River basin.

In the **Pirikita Alazani River basin** the Wurm glaciation was presented widely on the eastern slope of the Atsunda range and in the eastern part of the Pirikita (Tusheti) range. The hypsometrical indicators of these ranges mostly exceed $\sim$3800 m and they created favorable conditions for the existence of the glacial zone. Therefore, the glacial forms are well preserved in the relief in the form of cirques, modified troughs and moraines.

There were two glaciers in the Pirikita Alazani River headwater, which merged with each other. The **Tebulo glacier** is the major, the length of which was $\sim$11.0 km and ended at the height of $\sim$2400 m. It is difficult to identify the glacier extension boundaries by the morphological signs based on the field survey results. The typical glacial deposits are seen up to the height of ~2500 m; then there are fluvial-glacial deposits, but it is difficult to identify whether they belong to the Tebulo glacier or the glacier flowing down from the Ruani range. As the relief is built with slates and sandstones, it is easily washable and moraines cannot be preserved for a long time.

During the Wurm there was a ~5.5 km long of valley glacier on the eastern slope of the Amugho peak in the heads of the Larovanistskali River, the right tributary of the Pirikita Alazani River. The detailed decoding of the aerial images has made it clear that in Wurm this glacier descended to the Samruli River gorge, the left tributary of the river Tushetis Alazani. Many facts prove it, namely: the Samruli River gorge is cut in its head; there are degraded moraines in the middle stream of the river and the gorge has a shape of a modified trough. The analysis of the relief shows that the Samruli River headwaters were captured by the Larovanistskali River as a result of regressive erosion in the post-Wurm period.

The Pirikita (Tusheti) range is hypsometrically high from the Mount Kachu (3891 m) to the Mount Diklosmta (4285 m). In terms of old glaciation, it was the strongest center in the Pirikita Alazani basin. Modern and old glaciers in the Pirikita range were more widely presented on its northern slope than on the southern slope.

There were valley glaciers in the basins of the rivers located on the southern slope of the Pirikita range (the rivers of Khaoskhevi, Cheroskhevi, Chigoskhevi, Didkhevi, Khaiskhevi, and Parsmaskhevi). The length of the glaciers was ~5–7 km and their tongues did not reach down the bottom of the river Pirikita Alazani River gorge. The Khao glacier descended to the lowest point—to ~2140 m. The tongues of other glaciers ended at the height of ~2200–2400 m. The Wurm cirques are more or less well preserved in the heads of all valleys; their lower thresholds are located at ~2400–2500 m. The lower sections of the above-listed valleys are erosive and were not subject to glacial action.

During the Wurm, there were quite many cirque and valley glaciers in the Pirikita Alazani River basin.

The glaciation in the Wurm in the Eastern Caucasus had the properties similar to those of glaciers on the southern slope of the Central Caucasus at present.

6.5 Southern Georgia's Highland

Under the southern Georgia's highland we consider all those ranges and plateaus, which are located to the south of the intermountain plain: the ranges of Adjara-Imereti, Shavsheti, Arsiani, Trialeti, Samsari and Javakheti, and the Erusheti upland. In the elevated part of these orographic units there were small glaciers developed during the Late Pleistocene evidenced by the glacial forms remained there.

The mentioned ranges are completely studied by many scientists: P. Gamkrelidze, N. Kandelaki, B. Klopotovskiy, L. Maruashvili, D. Tsereteli, Sh. Tskhovrebashvili, G. Maisuradze, D. Tabidze, and others. All researchers indicate the presence of old glacial forms there. The only difference among them is in the scales of glaciation.

Our studies showed that in the southern Georgia's highland the Wurm glaciation has almost the same form as there is in the Eastern Caucasus today. In the Wurm, the firn line was located in the Adjara-Imereti range at the lowest altitude, at ~2200–2300 m, and in the Samsari and Javakheti ranges—at the highest altitude, at ~2500–2700 m. Neither of the ranges reached the glacial zone. However, the nival zone was presented quite clearly, where small cirque and valley glaciers were formed under the influence of the atmospheric precipitations. These glaciers were concentrated at the small sections of the ranges with their heights more than ~2400 m, for example: Sakornia, Mepistskaro, and Loboroti—in the Adjara-Imereti range; Karakaya and Mount Sakvelo—in the Trialeti range; Kanli Dagi—in the Arsiani range, and Abuli, Samsari, and others—in the Samsari range.

The degraded cirque forms remained there are of small sizes and accordingly, the glacier lengths were short—up to ~3.0 km. The largest glaciers were in the Samsari range. Valley glaciers were developed around the volcanic peaks of Abuli and Samsari; their maximum length did not exceed ∼5–5.5 km (Maruashvili 1961). The cirque forms start at the height of ∼2600–2700 m. The genesis of the small lakes at this location is associated with the glacial and periglacial processes. The forms typical to the nival zone, both of Wurm and post-Wurm, are well preserved in the Samsari and Javakheri ranges. The sediments of old rocky glaciers are a good evidence of this.

6.6 Holocene Glaciation

In contrast to a hypothesis of a stable Holocene climate, as indicated by oxygen isotope records from the Greenland Ice core (Johnsen et al. 1997) and northern Alpine lake sediments (von Grafenstein et al. 1999), a growing number of studies (e.g., Mayewski et al. 2004 and references therein) have demonstrated that distinct periods of climate change occurred repeatedly throughout the Holocene (Agatova et al. 2012). Mountain glaciers are potentially good climatic indicators, but most of the information on the dynamic responses of mountain glaciers to past climate changes comes from European glaciers which only form 3% of the world's glaciated area (Dyurgerov and Meier 1997). Knowledge of the Holocene glacier behavior of remote and inaccessible mountain regions, often characterized by a high degree of glaciation, is rather limited. For better understanding of the global spatial-temporal glacier variability and throughout the Holocene, it is very important to study worldwide glacier dynamics, including the Caucasus Mountains.

There were several stade glaciations in the Holocene and their paleoglacial study allows reconstructing the landscape zones and paleo-climate. With this purpose, the glacial deposits were studied on the southern slope of the Caucasus range and their correlation to the river terraces has been conducted. The absolute age (C^{14}) of organic deposits has been identified

at several locations. On the southern slope of the Caucasus the stade glaciation is studied by Maruashvili (1956), Tsereteli (1966), Kovalev (1961), Khazaradze (1968), Gobejishvili (1995), and others.

The trace of the stage glaciation in the Late Holocene (Neogene) is particularly well preserved in the **Kodori River basin** in the form of lateral and terminal moraines. The signs of an early times stade glaciation are survived in the Sakeni River gorge, while in other gorges their traces are washed out or they are in the form of the separate fragments making it difficult to identify their number and limits of expansion.

The first stade moraine in the **Sakeni River gorge** is at the height of ∼1550–1600 m, in the environs of its right tributary of Olbaki (Maruashvili 1961). At a distance of ∼1.5 km above it, there is another moraine, which is seen quite well in the relief.

At the heights of ∼1860 m and ∼1930 m, there are two moraine complexes in the Sakeni River gorge distanced by ∼500–600 m from each other. The upper (fourth) moraine of them is well seen in the relief.

After the mentioned stade glaciation there was a lake; after it was filled up, a straight, slightly inclined, and flat-bottomed valley was originated (Maruashvili 1961). The origination of the fourth stade moraine and the lake should have taken place at the beginning of the Late Holocene. Conditionally, we attributed the third moraine to the Upper Holocene. During this stade glaciation, the Sakeni and Achapara glaciers were merged with each other.

There are two stade moraines in the basins of almost all glaciers in the Sakeni River gorge. The distance between them is about 0.5 km. The young lateral and terminal moraines of them belong to the LIA maximum glaciation, and the intensely denuded moraine in the relief belongs to the historical stage (∼2000–2500 years) (Gobejishvili 1995).

In the Sakeni River gorge we see the traces of six stade glaciations. The I-III moraines of them were formed in the Late Pleistocene and Early

Holocene, and the IV-VI| stade moraines were originated during the Neo-glacial (Late Holocene) glaciation.

The stade moraines of the Late Holocene age are remained in the heads of the rivers of Klichi and Marukhi at the height more than ~1900 m above sea level. The IV stade moraine is located at the height of ~1900–1950 m; the V historical stade moraine is remained at the height of ~2150–2200 m and the VI LIA maximum moraine is remained at the height of ~2250–2300 m and is clearly seen in the relief. It is distanced from the glacier by ~1.0–1.5 km (Gobejishvili 1995).

The trace of stade glaciation in the Enguri River basin is best fixed in the gorges of the rivers of Mulkhura and Adishchala.

We distinguish the six stade glaciations in the **Mulkhura River gorge**. A full spectrum of the stade moraines is seen in the Mestiachala River gorge, on the western slope of Gvalda range, near the airport. The lateral stade moraines are fixed as steps, while the Wurm moraine is clearly seen as a hillock; the Wurm moraine is found as a hillock on the right side of the gorge, as well. The trace of (I-II) two stade glaciations is fixed as a step at the height of ~1750 and ~1810 m above sea level. Below that, at the height of ~1650 and ~1620 m, there are two more steps built with glacial material. The IV moraine of them is in the form of a hillock, which is stretched to the village of Spardishi. The moraine of the same stage is found at the village of Lanchvali, which is covered with delluvial cones. Lanchvali and Spardishi lateral moraines arch near the Lanchvali Bridge. The glacial boulders here evidence that during the IV stage, the glacier ended at this location. However, the terminal moraine is washed due to the river action (Gobejishvili et al. 2012).

The moraine material in the environs of a former touristic base in Mestia is a continuation of the III stade moraine, while its morphological continuation is found even higher, in the Mulkhura-Mestiachala flat watershed.

The traces of the I and II stade glaciations are remained on the left slope of the Mulkhura River, in front of Mestia-Lenjeri (Fig. 6.11). The II

stade moraine is presented as a hillock descending to the gorge bottom above Lenjeri. The trace of the I stade glacier is seen very slightly as individual steps and reaches the village of Kashveti. During these stages the Tviberi and Lekhziri glaciers were merged with each other.

The fragments of the V (historical stage) stade moraine are remained in the environs of the village of Lavladashi and as fragments on the left side up along the gorge. There are a number of granite boulders on the both sides of the gorge, both, in the floodplain environs and beyond it (Fig. 6.12). It should be noted that a morphologically well seen debris cone in front of Lavladashi gorge is of three-step (Fig. 6.13) identified through the correlation to the glaciation in the Neo-Glacial (Late Holocene) period. During the LIA maximum the Lekhziri and Chalaati glaciers did not merge with each other. The Chalaati glacier descended to the Lekhziri River bed and created the plug. The VI stade moraines are clearly seen in the relief.

The trace of the stade glaciation in the Mulkhura River basin is best remained in the environs of the Ughviri Pass and Lagvzalieri area. The relative height of the Ughviri Pass is 300–350 m from the Mulkhura gorge. During the stade glaciation (the I and II stages) the thickness of the Tviberi-Tsaneri glacier was more and one flow of the glacier (the Nageba glacier) moved to the Enguri River gorge. On the slope of the Enguri River, within the Wurm moraines, there are two series of lateral moraines and a corresponding terminal moraine. Such state of moraines evidences that the glacier flow of these stages did not reach the bottom of the Enguri River. On the northern slope of Ughviri two lateral stade moraines (III-IV) are identified. One of them is located immediately near the watershed and another one is located in the environs of the Ughviri Lake (the lake itself is originated during this stage).

On the right slope of the Mulkhura River gorge, at Lagvzalieri location, at the height of ~1900–1950 m, there is a ~1.0–1.5 km long moraine, which is located at the same hypsometrical height as Ughviri III and IV moraines. Even higher (~2120–2150 m), there is another

Fig. 6.11 The trace of the I and II stade glaciations on the left slope of the Mulkhura River in front of Mestia-Lenjeri (*photo by* L. Tielidze)

Fig. 6.12 Erratic boulder of the Mestiachala River (*photo by* L. Tielidze)

moraine, which in our opinion, is synchronous to the I and II stages of Ughviri. The fragments of the terminal moraines of the III and IV stages are found on the territory of Mestia community (at the beginning of the Mulkhura gorge), at the height of ~1450 m.

The V stade moraine is located at the confluence of the rivers of Tviberi and Tsaneri. This moraine is morphologically seen in the relief and without a doubt it was formed by the action of the Tviberi glacier. Corresponding to it the moraine is found in the Tsaneri gorge, in the environs of the mineral waters. These moraines belong to the historical stage. The LIA maximum moraine is completely presented in the basins of all glaciers.

Identification of the age of a stade moraine became possible only by correlating them to the river terraces. Gobejishvili (1995) sampled an organic substance (wood) from ~8 to 10 m high

Fig. 6.13 Lavladashi debris cone (*photo by* L. Tielidze)

terrace at the village of Jorkvali (the Enguri River gorge) with its age (∼ 5000 ± 200 years) identified at the Tallinn radiocarbon laboratory. This terrace seems to be formed at the end of the Middle Holocene. This terrace of the given height is fixed at many locations in the Enguri River gorge (villages of Pari, Tskhumari, Lenjeri, etc.) and ends at the IV stade moraine near the settlement of Mestia.

There is a 2–3 m high terrace found at Mestia and at the village of Lenjeri, which stretches long the gorges of the rivers of Mestiachala and Mulkhura. In the Mestiachala River gorge an airport and a new settlement are built on this terrace, which is morphologically clearly seen and stretches to the village of Lavladashi, where it is transformed into the historical stade (V) moraine. In the Mulkhura River gorge this low terrace is seen everywhere and is transformed into the stade moraine at the confluence of the rivers of Tviberi and Tsaneri.

The analysis of the materials allows making the following conclusion: the IV-V and VI stade moraines were originated in the Late Holocene (Neo-glacial period).

The **Adishchala River gorge** beyond the village of Adishi is of a trough shape. The bottom of the valley is slightly inclined and is of ∼ 200 m long from the village to the glacier. The glacial deposits are found on the right side of the gorge and cover entire slope. A trough of the Wurm age and its arm is well seen on the right side of the gorge. There are steps at different

locations of the trough slope, which get lower along the gorge and reach down its bottom. It is possible to identify the boundaries of stade glaciations by these steps.

The tongue of the modern Adishi glacier is clean and the lateral moraines have no great thickness. Generally, Adishi glacier is characterized by the lack of the moraine material, unlike other large glaciers. Similar picture was observed during other stade glaciations as well. That is why there are steps remained due to the action of the glacier and not the lateral moraines on the slope. As for the terminal moraines, they either are covered by debris cones, or are strongly denuded. On the left slope of the gorge there is no glacial material remained there due to the active erosive-denudation processes.

Despite this difficulty, we have identified the trace of six stade glaciations in the Adishchala River gorge. The I and II stage moraines are remained at the village of Adishi and at the former farms of Adishi (at the heights of ∼ 2050 and ∼ 2100 m). Their corresponding steps are seen on the slope and these steps descend to the gorge bottom at these locations. The signs of the next stade glaciation are found up to the church of Lichanishi (III) and its environs (IV). To our mind, the terminal moraines are covered with the material of Lichanishi cone. The evidence is that the section of the glacier gets under the debris cone. Beyond the church, on the left slope of the gorge, there are steps slightly seen as a fragment, which are built with glacial material (granites).

Only the LIA maximum moraines are well seen in the Adishchala gorge. There is an intensely denuded historical stade lateral moraine at a distance of ~1.0 km below them.

The trace of a stade glaciation in the Adishchala gorge is well preserved in the heads of its right tributaries (Lichanishi and Tvibi), where today and in the past, there were developed the rocky glaciers. There are four steps of the rocky glaciers and two lateral moraine hillocks in the Lichanishi gorge, and there are five steps of the rocky glaciers and one moraine hillock in the heads of the Tvibi River. Three steps well seen in the relief were originated in the Late Holocene (Gobejishvili, Rekhviashvili 1977, 1986, 1988) by the rocky glaciers.

The trace of the stade glaciation in the gorges of other rivers of the Enguri River basin is seen more or less clearly.

The trace of the stade glaciation in the Rioni River basin is well preserved that caused the conduction of complete studies.

The trace of the stade glaciation in the **Bubistskali River basin** is entirely preserved as lateral and terminal moraines. On the right side of the gorge, on the bottom of the Wurm trough, near the site of Tbaukebi, there are six moraine hillocks, which are presented in pairs in the relief.

During the I stade glaciation, the Buba glacier descended almost to Shovi (~1500 m). This is evidenced by the morphology of the gorge and glacial boulder material on the territory of Shovi, which is presented as granites. Kuznetsov (1931)

and Tsereteli (1966) talked about the presence of the stade moraines at this location even earlier.

The terminal moraines of the II and III stages are washed in the gorge. Despite this, identification of the boundaries of these stade glaciers is possible based on the morphology and direction of the lateral moraines. Buba glacier descended to the height of ~1700 m in the second stage and to the height of ~1900 m in the third stage.

The trace of the IV stade glaciation is seen as the lateral moraines on the both sides of the gorge, which arch at ~2000 m and are transformed in a terminal moraine.

The trace of the V stade glaciation is denuded and is remained as a fragment. Its corresponding terminal moraine is found at the height of ~2200 m.

The sixth moraine belongs to the most recent (LIA) stade glaciation and is seen in all basins.

Based on the conducted surveys we identified the lengths and firn line depression of the Buba glacier during the stade glaciation (Table 6.6).

The terminal moraines in the **Chveshura River gorge** are well preserved near the village of Gona, at the altitude of ~1750 m above sea level. ~250 m above the moraine the archeologists found a copper mine. The enriching of the materials mined from there was made on the moraine surface immediately, evidenced by the bulk of archeological materials. According to the radiocarbon method the mine of Gona dates back to the ~16–15th cc. B.C. (Gobejishvili 1995).

There is a next stade moraine at the confluence of the rivers of Chveshura and Dombura

Table 6.6 Lengths and firn line depression of the Buba glacier during the stade glaciation	Stage	Length, ~km	Firn line depression, ~m
	Modern	3.0	–
	VI	4.5	180
	V	5.5	270
	IV	8.5	450
	III	9.0	480
	II	11.0	590
	I	13.5	620

at the height of ~2050–2100 m; and the LIA moraine is fixed in ~1.5 km from the glacier.

The detailed surveys conducted in the basin of the Bubistskali River basin have allowed us to generalize these materials and apply them to the **Zopkhitura River basin**.

The trace of the Holocene stade glaciation is well survived in the relief. The number of the moraines is six and they are distributed in pairs. The moraines are remained on the left slope. The boulder material is distinguished for its large sizes and is slightly processed. The distance between the moraines is ~0.8–1.0 km. The carbonate flysch deposits building the relief between the moraine layers are outcroped. The III and the IV stade moraines are located in the environs of the Veligrdzela location. The fragments of the III moraine are located on the left slope. At a distance of ~1.0 km from the III moraine there is the IV moraine. The strong delluvial material in the environs of Khirkha covers the moraines of this stage. The fragments of the V moraine are remained at the height of ~2050 m as a hillock built with granites on the left side of the valley. As for the VI stade moraine, it is survived at the height of ~2120 m.

The trace of the stade glaciation in Late Holocene is well remained in the basins of the rivers of Notsarula, Gharula, Jejora, Ghobishuri, Latashuri, Ritseuli, Koruldashi, Laskadura, and other basins.

Based on the conducted surveys we identified the lengths of the glaciers and firn line depression in Late Holocene for individual stages.

During the IV stage (~4000–4500 years ago) the length of the valley glaciers in the Rioni River basin was ~8–10 km and the length of the cirque glaciers was ~5–6 km. The firn line depression was ~450–470 m (Gobejishvili 1981).

During the V stage (~2000–2500 years) the length of the valley glaciers was ~5–6 km and that of the cirque glaciers was ~2.5–3.0 km. The firn line depression was ~270–300 m.

During the LIA stage the length of the valley glaciers was ~3–5 km and that of the cirque glaciers was ~1.5–2.0 km. The firn line depression was ~150–180 m.

References

Abich H (1865) Isledovanie nastaiashix i drevnix led-nikov Kavkaza (The study of current and ancient glaciers of the Caucasus). Cb, cvedenii o Kavkaze. T. 1. Tiflis (in Russian)

Agatova AR, Nazarov AN, Nepop RK, Rodnight H (2012) Holocene glacier fluctuations and climate changes in the southeastern part of the Russian Altai (South Siberia) based on a radiocarbon chronology. Quatern Sci Rev 43:74–93

Astakhov NE (1973) Structural Geomorphology of Georgia. Pub. House "Metsniereba", Tbilisi (in Russian)

Dondua G (1959) Jejorisa da Garulas auzebis geomor-fologia (Geomorphology of Jejora nad Gharula river basins). Vaxushtis saxelobis geografiis isntitutis shro-mebi, t. 12 (in Georgian)

Dumitrashko NV, Lilienberg LA, Antonov BA, Baljan SP (1961) Ancient glaciations of the Caucasus aud their correlation with glaciations of the Russian plain. Trudi komisii po izucheniju chetvertichnogo perioda 19:170–180 (in Russian)

Dyurgerov MB, Meier MF (1997) Year-to-year fluctua-tions of global mass balance of small glaciers and their contribution to sea-level changes. Arct Antarct Alp Res 29(4):392–402

Ehlers J, Gibbard P (2008) Extent and chronology of Quaternary glaciation. Episodes 31(2)

Gobejishvili RG (1995) The evolution of the modern ice age glaciers and mountains of Eurasia in the Late Pleistocene and Holocene. The thesis of doctor of science degree in geography (in Georgian)

Gobejishvili RG, Tielidze L (2012) The map of the modern and Late Pleistocene (Wurmian) glaciations of Georgia. National Atlas of Georgia. Publishing house "Cartography", Tbilisi, p 91 (in Georgian)

Gobejishvili R, Lomidze N, Tielidze L (2011) Late Pleistocene (Wurmian) glaciations of the Caucasus. In: Ehlers J, Gibbard PL, Hughes PD (eds) Quaternary glaciations: extent and chronology. Elsevier, Amster-dam, pp 141–147. doi:10.1016/B978-0-444-53447-7. 00012-X

Gobejishvili RG, Tielidze L, Lomidze N, Javakhishvili A (2012) Monitoring of the glaciers against the back-ground of the climate change. Publishing House "Universali", Tbilisi (in Georgian)

Johnsen SJ, Clausen HB, Dansgaard W, Gundestrup NS, Hammer CU, Andersen U, Andersen KK, Hvid-berg CS, DahlJensen D, Steffensen JP, Shoji H, Sveinbjornsdottir AE, White J, Jouzel J, Fisher D (1997) The delta O-18 record along the Greenland Ice Core Project deep ice core and the problem of possible Eemian climatic instability. J Geophys Res Oceans 102:26397–26410

Khazaradze RD (1968) Old Glaciation of the South Slope of the Great Caucasus. Metsniereba, Tbilisi, 135 p (in Russian)

Khazaradze RD (1971) Relief, kontinentalnie otlojenia i pleictocenovie oledenenie bac. R, Inguri (Relief, continental sediments and Pleistocene glaciation Enguri River basin). Avtoreferat kand. Disertacii. Tbilisi (in Russian)

Klopotovskiy BA (1949) Xorlakelskaia morena (Khorlakelskaya moraine). Soobsh. AN. GSSR. T. 10, N. 6 (in Russian)

Kovalev PV (1961) Sovremennie i drevnie oledenenie bacceina r. Inguri (Modern and ancient glaciation Enguri River basin). Mat. Kavkaz. Eksped. T. 2. Xarkov. Izd XGU (in Russian)

Kuznetsov IG (1931) Geoologicheskie stroenie kurorta shamshovi c bassein reka chanchaxi v centr, Kavkaze (Geological structure Shamshovy resort to the Chanchakhi river basin). tr. vsesoiuznogo geolog, razvitia obedinenia. vipusk 151 (in Russian)

Maruashvili LI (1956) Advisability of revising existing paleogeographic conditions of the glaciation age in the Caucasus. Academy of Sciences of the Georgian SSR, Tbilisi (in Russian)

Mayewski PA, Rohling EE, Curt Stager J, Karlen W, Maasch KA, David Meeker L, Meyerson EA, Gasse F, van Kreveld S, Holmgren KU (2004) Holocene climate variability. Quatern Res 62:243–255

Milanovskiy EE (1960) O sledax verxnepleceinnogo oledinenia v visokogornoi chasti centralnogo kavkaza (Traces Late Pleistocene glaciation in the high part of the Central Caucasus). Dokladi an SSSR. T. 130, N. 1 (in Russian)

Nemanishvili S (1962) Kvemo rachis chrdiloet nawilis geomorfologiisatvis (Geomorphology of the northern part of Lower Racha). Vakhushtis saxelobis geogr. institutis shroebi. T. 12 (in Georgian)

Paffenholz KN (1958) Novie daannie po ctratografii lav Kazbegskogo raiona i kelskogo vulkanicheskogo plato (Centralnii Kavkaz) i drevnomu oledeneniu etoi oblasti (The new stratografic data of Kazbegi district and the Keli volcanic plateau (Central Caucasus), and ancient glaciation in this area). Sovetskaia geologia, N. 12 (in Russian)

Penck A, Bruckner E (1901/1909) Die Alpen im Eiszeitalter, 3 vols. Tauchnitz, Leipzig, 1199 p

Reingard AL (1937) About the problem of the stratigraphy of the Glacial Time of the Caucasus. In: Transactions of international INQUA congress transactions. Section M 1, pp 9–31 (in Russian)

Reinhardt AL (1936) Lednikovii period kavkaza i ego otnoshenie k oledeneniu alp i altai (Ice Age of the Caucasus and its relation to the glaciation of the Alps and the Altai). tr. II. mejd. konf. po izuchenie chetvertichnogo perioda evropi. L-M, bip. 2 (in Russian)

Serebruanny LP, Orlov AV (1989) Moraines as the source of glacial information, 235 p (Unpublished manuscript)

Tabidze DD (1965) Gomorfologia baseina reka Kodori (Kodori river basin geomorphology), Avtoreferat kand. Disertacii, Tbilisi (in Russian)

Tsereteli DV (1959) Izmenenie lednikov iuzhnovo sklona Kavakiona za poslednie 25 let (Changing the southern slope of the Caucasus glaciers over the past 25 years). Coobsh. An. GSSR. T. 21. #6. Tbilisi (in Russian)

Tsereteli DV (1966) Pleistocene deposits of Georgia. Publishing House "Metsniereba", Tbilisi, 582 p (in Russian)

Vardaniants LA (1930) O novom sposobe podcheta depresii snegovoi granici v sviazi izucheniem stadii otstupania lednikov gornoi gruppi adai-xox v centralnom kavkaze (On a new method estimate depression snow line in relation the study stage glacier retreat mining group Adai-hoch in the Central Caucasus). Izd. RGO. T. 62. Vip. 2 (in Russian)

Vardaniants LA (1937) The ancient glaciation of the rivers Iraf (Urukh) and Tsey (Central Caucasus). Proc State Geogr Soc 69(4):537–562 (in Russian)

Von Grafenstein U, Erlenkeuser H, Brauer A, Jouzel J, Johnsen SJ (1999) A mid-European decadal isotope-climate record from 15,500 to 5000 years BP. Science 284:1654–1657

Tsereteli DV (1959) Izmenenie lednikov iuzhnovo sklona Kavakiona za poslednie 25 let (Changing the southern slope of the Caucasus glaciers over the past 25 years). Coobsh. An. GSSR. T. 21. #6. Tbilisi (in Russian)

Tsereteli DV (1959) Izmenenie lednikov iuzhnovo sklona Kavakiona za poslednie 25 let (Changing the southern slope of the Caucasus glaciers over the past 25 years). Coobsh. An. GSSR. T. 21. #6. Tbilisi (in Russian)

Bibliography

Freshfield DW (1896) The exploration of The Caucasus, Vol. II, Edinburgh: T. and A. Constable, printers to her majesty, London and New York

Printed by Printforce, the Netherlands